CW01329328

AIR BATTLES IN MINIATURE
A Wargamers' Guide to Aerial Combat 1939-1945

AIR BATTLES IN MINIATURE

A Wargamers' Guide to Aerial Combat 1939-1945

Mike Spick
Foreword by Alfred Price

Patrick Stephens, Cambridge

© Mike Spick 1978

All rights reserved.
No part of this publication may be reproduced,
stored in a retrieval system or transmitted,
in any form or by any means, electronic,
mechanical, photocopying, recording or otherwise,
without prior permission in writing
from Patrick Stephens Limited

First published—1978

British Library Cataloguing in Publication Data
Spick, Mike
    Air battles in miniature
    1. War games, Air
    I. Title
    793'.9   U310
    ISBN 0 85059 296 8

Text photoset in 10 on 11pt English 49
by Stevenage Printing Limited, Stevenage.
Printed in Great Britain on 100 gsm
Buccaneer Matt and bound by
The Garden City Press Limited, Letchworth
for the publishers, Patrick Stephens Limited,
Bar Hill, Cambridge, CB3 8EL, England.

Contents

Foreword by Alfred Price	7
Introduction	8
Chapter 1 — In the beginning	
A brief look at pre-war outlooks, attitudes and developments	9
Chapter 2 — Flight	
Elementary principles of flight and propulsion . . . the composition of a combat aircraft . . . weapons	12
Chapter 3 — Combat flying, aerobatics and tactics	
Defensive/offensive . . . strategic/tactical . . . fighter aircraft requirements and tactics . . . the use and non-use of aerobatics . . . relative speeds . . . the effect of altitude on performance	18
Chapter 4 — Compromise and the needs of wargaming	
Model scale . . . distance scale . . . the third dimension . . . move time . . . calculation of speed . . . calculation of rate of turn . . . turning device . . . level turns . . . acceleration . . . deceleration . . . calculation of climb rates . . . calculation of dive rates . . . stalling . . . spinning . . . cruising speed . . . maximum level speed . . . terminal velocity . . . the effect of external loads on performance . . . how to calculate performance data on a 'special'	23
Chapter 5 — Observation	
The clock code . . . probabilities of spotting for single-seaters . . . adjustment for multi-seaters . . . the sun . . . cloud cover	40
Chapter 6 — Aerial gunnery	
Comparison of gun performance . . . evaluation of firepower . . . calculation of fire factors . . . ammunition supply . . . calculation of recoil effect . . . assessment of hit probability . . . evaluation of fire effect . . . calculation of damage effect . . . air-to-air gunnery . . . gunsights . . . range . . . deflection . . . effect of relative speed . . . calculation of relative speed . . . air-to-air bombing . . . air-to-air rocketry	44
Chapter 7 — Air-to-ground and ground-to-air	
Types of bombs . . . bomb sights . . . level bombing . . . glide-bombing . . . dive-bombing . . . anti-aircraft guns . . . ground-to-air gunnery . . . conversion of the air game to tie in with surface battles . . . mechanics of explosions . . . radius of bomb effect . . . level bombing error . . . level bombing result circles . . . dive-bombing error . . . fighter-bomber bombing error . . . low-level bombing attacks . . . air-to-ground rocketry . . . air-to-	

ground gunnery ... categories of ground targets ... structures ... vehicles ... infantry ... effect of bombs, rockets and guns against ground targets ... napalm ... air-to-ground observation ... ground-to-air observation ... anti-shipping strikes ... skip-bombing ... ship classification ... effect of bombs on ships ... torpedoes ... torpedo tactics ... torpedo effect ... rocket attacks on shipping ... depth charges ... ship-to-air gunnery ... anti-aircraft fire screen ... screen values   **61**

Chapter 8 — The table-top game
Setting up ... aircraft performance data ... playing rules ... scenario suggestions ... solo game suggestions   **91**

Chapter 9 — Large-scale air warfare
Appreciation of the problems ... scenario suggestions ... the 'ops-room' table principle ... map making ... preparation of an 'ops-room' table playing surface ... weather ... airfield unserviceability ... plotting and tracking ... radar capability ... counters ... move time ... take-off ... climb rates ... speeds ... calculation of fuel consumption ... fuel units ... raid planning ... forecast wind ... navigational errors ... observation ... air-to-air radio ... interception ... dogfight ... surprise ... firing ... morale ... bombing effect ... airfield attack ... sea searches ... carrier strikes   **123**

Chapter 10 — Campaign details
Battle of Britain/Malta 1942 ... general considerations ... countering hindsight ... organisation and serviceability ... order of battle ... redeployment ... turnround times ... replacements ... targets ... special considerations for Malta   **141**

Chapter 11 — Simulating night fighting
Problems ... aids for night bombing ... countermeasures ... airborne radar ... ground radar ... jamming and other rotten tricks ... general progress ... appraisal of game requirements ... scenario suggestions ... move/time scale ... movement ... electronic detection ... interception ... competition game suggestions   **151**

Select bibliography   **160**

Foreword by Alfred Price

My association with Mike Spick began back in 1975, soon after the publication of my book *World War II Fighter Conflict*. Out of the blue he rang me, and asked whether I would be interested in watching an air combat wargame he had put together based partly on the information given in my book. Now up to that time I had had a rather fuzzy pre-conceived notion that wargamers were grown-ups who played around with kids' toys, and tried to make out that they were making some serious contribution to military understanding in the process. More out of curiosity than anything else, I went along to see Mike's game. I am glad that I did. It turned out to be a stimulating and enjoyable evening, during which I was greatly impressed by Mike's ability to grasp the essentials of air combat and apply them to his simulation.

Of course, no wargame can ever be just like the real thing. The only thing exactly like fighting against a Messerschmitt 109 with a Spitfire I in 1940 is fighting against a Messerschmitt 109 with a Spitfire I in 1940. Mike is the first to admit that he has had to make several departures from reality, in order to make it possible for the average wargamer to run the simulation on a dining-room table. Nevertheless, his air combat simulation probably comes as close as any game can to giving the 'feel' of what air combat is really like.

Air Battles in Miniature is Mike Spick's first book and in it he demonstrates a competent grasp of his subject. I wish this work the success it richly deserves.

Alfred Price
January 1978

Introduction

My purpose in writing this book is to fill a large gap. The most popular wargaming periods are the most spectacular; Ancient and Napoleonic. Why then are the air battles of World War 2, surely the greatest spectacles in the entire history of warfare, so neglected? Granted, they are difficult to reproduce, but far from impossible.

My initial aim was to produce a handbook on air wargaming. It is never enough to produce a set of playing rules which say 'do this'. I have therefore tried to take the reader step by step through the analysis of data both on aircraft performance and weaponry and explain in detail how the playing rules were formulated, but without becoming involved in either advanced aerodynamics or mathematics. No less than three sets of playing rules are included. The first covers the table-top game played with model aircraft. While primarily concerned with the fighter-versus-fighter game, it also covers bombing, ground attack and anti-shipping strikes for the benefit of those who wish to bring a greater degree of realism into their land or naval battles. The second game covered is the large-scale air campaign, which is fought on the 'ops-room table' principle. Sufficient campaign details are given for the reader to re-create either the Battle of Britain or Malta, 1942. Finally we have night fighting, which is a do-it-yourself board game, and no less effective for that.

Many people have assisted in the preparation of this book. In particular I should like to thank Alfred Price, many of whose suggestions are incorporated and on whose specialised knowledge I have leaned heavily; Christopher Shores, whose knowledge of the Middle East air war is unrivalled; Edward Sims for permission to quote from *The Fighter Pilots*; Keith Robinson for assistance on naval affairs; Dave Smith of Deeping Models for producing the aircraft shown in the action sequences; Brian Monaghan for some excellent model photography; Martin Holbrook for the maps and diagrams; Stuart Pudney for checking the original draft; and the staff of the Imperial War Museum at Duxford for their unfailing assistance. Finally a big vote of thanks to Minyon Prescott, who decoded my scrawl, typed the manuscript, and bore without complaint all my amendments.

To produce a definitive book on wargaming would be presumptuous; if I have succeeded in laying a solid foundation, then I am satisfied.

Mike Spick
Market Deeping, January 1978

Chapter 1

In the beginning

The mid-1930s were a period of transition for martial aviation. Until then, the design of fighter aircraft had been influenced mainly by the hard-won experience of the Great War, supported by too many showings of 'Hells Angels'. During the later stages of this conflict, aircraft performance did not differ widely between the nations in such matters as top speed and rate of climb. Add to this the fact that most aerial combat took place relatively close to the trenches with the main object of attaining local air superiority over the battlefield, chuck in large numbers such as the Richthofen Circus, and the result was similar to Piccadilly in the rush hour, with dozens of aircraft milling about in the space of less than half a cubic mile. Given, these circumstances, survival of the individual was dependent on two factors: luck and ability. Luck needs no further explanation, but pilot ability consists of experience and aircraft handling. Experience, of course, depended on surviving and was a factor which everyone started without, and gained the hard way. The ability to handle an aircraft better than one's opponent was, therefore, paramount, and this was conditioned by the 'flyability' of the aircraft. It was no good being a better flyer than the nasty man on one's tail if his aircraft was to any marked degree superior. In this situation you could only hope that someone on your side had aggressive intentions towards him, or that he was a rotten shot, or preferably both. Thus manoeuvrability came to be generally accepted as the prime requirement of a fighter aircraft. I think I am right in saying that the only major exception to this rule was the SE5, but then there are always exceptions to every rule.

This design philosophy continued well into the 1930s. The fighter pilot was always invested with glamour, partly built on the exploits of the war when, in contrast to the mud and carnage of the trenches, his exploits in the clean blue sky seemed to savour more of the colour and pageantry of mediaeval chivalry. This chivalric tradition was perpetuated between the wars at air displays, when the knights of the air, clad in silver fabric and blazoned with squadron markings and crests, swooped and sparkled in the sky, their engines snarling out a song of battle worthy to compare with the Valkyries of old. Heroic then they seemed, unconquerable. To fly was the greatest adventure possible, and young men became service pilots through love of flying, with scarcely a thought of fighting. The arena of the air took on the aspect of a giant tourney, where gaily caparisoned champions met and jousted before going home to tea. Aerobatic flying was the finest form of flight, the most enjoyable, and this attitude helped continue the trend towards manoeuvrability, which was further aided by the assumption that any future conflict would see massed dogfights on the old lines, with the ability to stay behind an opponent or throw off an attacker as the prime requirement. The most extreme nation in this respect was Italy, whose fliers were vigorously objecting to such newfangled ideas as monoplanes and enclosed cockpits even after war had commenced. Splendid aerobatic pilots, the Italians had taken their attitudes to such an extreme that they even held world records for sustained inverted flying which, unless they were planning an invasion of Australia, could have been of no possible practical use.

At this stage there arrived the flying banana skin, otherwise known as the high speed monoplane bomber. This gave rise to a situation where the fighter was unable to catch the bomber except in a dive, and even if able to catch it was unable to stay with it for long enough to shoot it down. The writings of Douhet on strategic bombing had come and gone largely unremarked except by the few, but the dictum 'the bomber will always get through' took on a new significance. The immediate counter was the development of the high speed fighter aircraft, which was of necessity a monoplane. This configuration, coupled with such newfangled ideas as retractable undercarriages and enclosed cockpits, gave greater aerodynamic cleanliness and, therefore, greater speed than the biplane, but at the expense of manoeuvrability and rate of climb. (The same power/weight ratio would always give the biplane a better rate of climb in comparison with the monoplane.) The general consensus in most air arms was that the function of the fighter was to destroy the bomber. Fighter combat was looked on as a thing of the past as it was thought that the centrifugal stresses imposed on the pilot by high speed manoeuvres would render dogfighting impossible, added to which the short length of time that the sights could be brought to bear would negate effective shooting. The general concept was similar to that of the recipe for jugged hare. It starts 'first, catch your hare'.

Another fighter concept arose in the mid-1930s; that of the long-range heavy fighter. This had of necessity to be twin-engined, and the extra weight and consequent lack of agility were thought not to matter as the days of the dogfight were no more. Its function was to carve its way through a defending fighter force and protect the bombers it was built to escort, as much by giving the defence problems as by any inherent fighting ability. This gave birth to the Bf 110 for Germany, the P-38 for the USA and the Fokker G1 in Holland. The Royal Air Force had no equivalent, its own heavy fighter force being makeshift; the Blenheim bomber being converted, and the Beaufighter owing much of its design to the Beaufort. The RAF considered that its strategic bomber force, armed with multi-gunned power-operated turrets, flying wingtip to wingtip, could concentrate enough firepower to discourage any attacker. They did, however, produce a very odd bird in the Defiant, a single-engined two-seater, armed with a four-gun power-operated turret, the function of which was to formate with an enemy bomber force and knock hell out of it. The Americans, at about the same time, produced the Airacuda, which was a combination of both the Defiant and the heavy fighter, but fortunately this never saw service.

Tactical thinking during this period was distinctly muddled. The defence of the United Kingdom was prepared on the basis that air attack would only come from Germany, across the North Sea, by unescorted bombers. This premise having been accepted, RAF squadrons flew in tight formation, in 'vics' of three aircraft, presumably because it looked pretty at air displays. Only the leader could keep an effective lookout as the rest were preoccupied in keeping station. To guard against

In the beginning

attack from the rear, a pair of 'weavers' were used, usually the most junior pilots on the squadron. This practice was discontinued in 1940 due to a severe shortage of weavers. To further demonstrate the British sporting spirit, a formal system of fighter attacks was introduced. In practice this meant that not too many enemy aircraft were fired on at once, also that enemy gunners were not confused by having an excessive number of targets. The one unsporting innovation at this time was the introduction of the eight-gun fighter, and, of course, radar. But more of these later.

Of the main combatant nations in World War 2, some had been distinctly unsporting. Britain, France and Germany entered the war in September 1939, Italy in mid-1940, Russia in mid-1941, Japan and the USA in December 1941. Prior to the outbreak of war, a dress rehearsal had been held in Spain, the participants being Germany, Italy and Russia. The Luftwaffe practised its role of flying artillery assiduously, also learning that fighter combat was possible after all; it demanded new tactics which were developed and used. In short, it learned a lot, which is more than can be said for the Italians and Russians. A similar but larger scale dust-up went on in China at the same time, with the Japanese learning the wrong lessons, due to two main factors. The first was that the Chinese were flying everyone else's cast-offs, and could by no means be classified as a modern air force. The second factor was the exaggerated number of 'kills' claimed in air combat, which had the effect of making the Japanese rest on their laurels of technical and racial superiority. Even the entry of Russia into the conflict for a few months did nothing to dispel the general complacency. Russia, feeling that she had had rather the best of it, also seemed to learn little or nothing. In fact, the Russians, having, with the Polikarpov I 16, introduced the world's first modern monoplane fighter, appear to have concluded that the ideal combination was a mixed fighter force comprising fast monoplanes which could catch anything trying to escape, combined with nimble biplanes to outfly the enemy when caught. However, it would be as well to have a brief look at the problems of air fighting before going further.

Chapter 2

Flight

Man not having been born with wings, flying is an unnatural activity, and is only to be undertaken with suitable mechanical aids. The first essential is wings, which is why aeroplanes have them. The second, and equally important essential, is a source of power. Power is converted into motion, and motion in turn is converted into lift, by the flow of moving air over the wing which is so designed as to induce a low pressure area on the top surface—a form of suction—and a high pressure area on the underside, a push upwards.

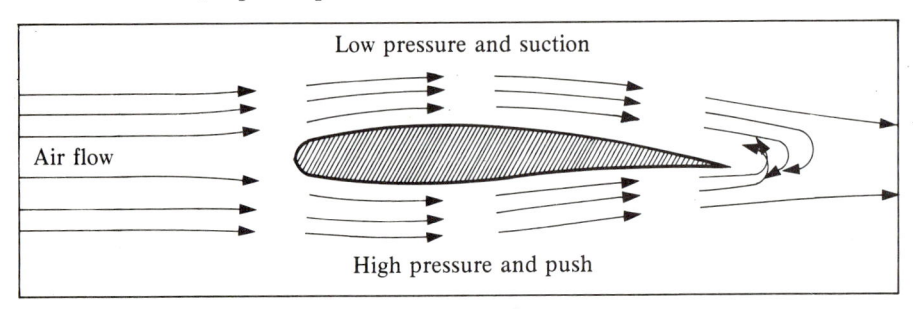

[Diagram of airfoil showing: Low pressure and suction above, High pressure and push below, with Air flow arrows]

   This, if the designer has calculated correctly, will give sufficient lift at speed to raise the aircraft from the ground. There are, however, other factors involved. The wing has to push air out of the way, and the faster the motion, the more resistance, and consequently more power is needed to overcome it. The air moving over the surface of the wing causes friction, which also produces a slowing effect. These two effects are called drag. The third effect is at the trailing edge of the wing where the low pressure area above meets the high pressure area below, aggravated by the need for new air to fill the space displayed by the moving wing. This causes all sorts of eddies and vortices to form, and is called turbulence. Any projection from the aircraft will cause turbulence, as will the aircraft itself. An aircraft designed to produce as little drag and turbulence as possible is called aerodynamically clean, and can thus travel faster for a given power output than a less clean aircraft.
   The next thing to consider is the source of power. The most common power unit used between 1939 and 1945 was the internal combustion engine. Other systems were used and we shall consider these briefly, but the internal combustion engine was by far the most commonly used source of power. Two basic types were in use: first was the in-line liquid-cooled engine, which had its cylinders arranged in either two rows, usually in a 'V' shape, or in four rows in either 'H' or 'X' layout. This

system made for a much more streamlined shape and greater cleanness in design, but suffered a weight penalty through having to carry coolant, radiators and the associated plumbing. The coolant system itself was very vulnerable to battle damage as, if any part of the system was ruptured and the coolant escaped, the engine would overheat and seize up very quickly.

The other basic type was the radial engine, in which the cylinders were arranged in one, two, or more banks, radiating outwards like the spokes of a wheel. It was cooled by the airflow past the cylinders in the same way as a modern motorcycle but, being dependent on the airflow, the engine temperature was more dependent on factors such as climatic conditions than the in-line, and consequently had a tendency to overheat, the classic example being the Brewster Buffalo, which was a success in the Russo-Finnish conflict and a flop elsewhere. The virtues and vices of the radial engine were the complete reverse of the in-line. It lacked aerodynamic cleanness, but could absorb more damage due to having no cooling system.

The internal combustion engine transmitted its power by means of a propeller, the blades of which were so shaped that they flung air backwards, the reaction to this tending to push the propeller forwards. As the propeller was in most cases securely fastened to the aircraft, this meant that the aircraft also moved forwards. The trouble was that the displaced air did not travel straight backwards, but more like a horizontal whirlwind. Consequently the swirl of air past the aircraft tended to screw it in the opposite direction to the propeller. This effect is known as torque, and explains why single-engined fighters tended to be able to turn faster to one side than the other. Normally the aircraft had to be trimmed to counteract this effect, which could be rather worrying at take-off with the engine producing full power; the aircraft tending to drift to one side and fly with one wing low. With twin-engined aircraft this could be overcome by having the propellers rotate in opposite directions, but as the thrust line for each engine was asymmetric, the problem was not as great. Towards the end of the war, several types of aircraft were flying with contra-rotating propellers; that is, two propellers rotating around the same shaft but in opposite directions. This involved a weight penalty for the gearing system, but it did reduce the torque problem.

The amount of air shovelled backwards by the moving propeller is dependent on the pitch, or blade angle. In coarse pitch the blade angle is quite obtuse and scoops a greater volume of air per revolution than fine pitch, in which the angle is acute.

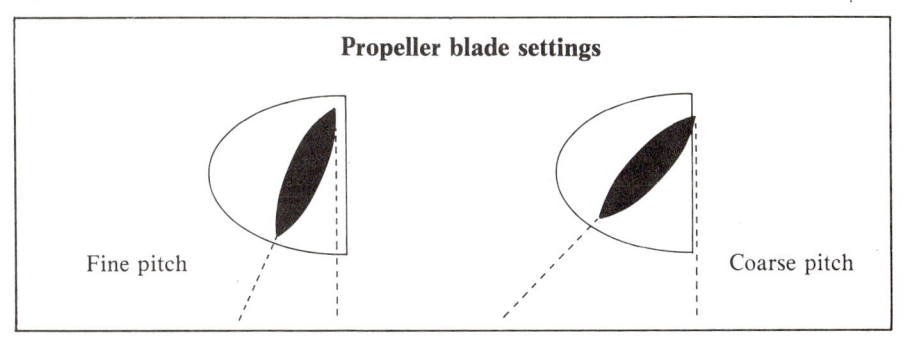

Propeller blade settings

Fine pitch   Coarse pitch

Fine pitch, moving less air per revolution and, therefore, doing less work, would enable engine revolutions to build up much faster, thereby giving better acceleration. Fine pitch would also be used during the take-off. Coarse pitch would be used once the need for rapid acceleration was over, as moving a greater quantity of air per revolution, it would maintain a higher air speed per given engine speed than fine pitch. It would also reduce fuel consumption for a given air speed. All the foregoing is, however, an oversimplification, as the density of the atmosphere reduces at height, thereby altering the optimum pitch required. The RAF started the war with fixed-pitch propellers which were neither one thing or the other, rapidly converting to two-pitch propellers which were rather better at the extreme ends of the performance envelope but were only a compromise in between, the pitch being set manually by the pilot. Then, at the height of the Battle of Britain, final conversion to hydraulically operated infinitely variable-pitch propellers took place which adjusted automatically to meet the needs of the moment, but could still be manually over-ridden by the pilot at will. The Bf 109E started out with a variable-pitch prop which was controlled by the pilot. This was superior to the British two-pitch prop but inferior to the British variable-pitch prop, chiefly in that it gave the pilot more work, and was rather demanding on his judgement at times.

The final point before leaving the subject of propellers is size. As engines were developed to give more power, a bigger propeller was needed to absorb it and convert it into thrust. At first this was overcome by increasing the number of blades, which increased from two to five during the period of hostilities. In addition the size had to be increased, needing greater ground clearance with correspondingly longer undercarriages. Probably the greatest effect this had on design was in the F4U Corsair, which had its wings cranked downwards to reduce the length of 'leg'.

Having now arrived at a suitable power plant and wings capable of taking to the air, we come to the thing which, in addition to holding them together, provides space for the driver and all the other little odds and ends, namely the fuselage, which also keeps the tail away from the wings. It is the aviation equivalent of a ladies' handbag, in that it would be unnecessary but for the need to house the radio, oxygen bottles, various tanks full of inflammable liquids together with their associated plumbing, crew members naturally and, in larger aircraft, even such things as bombs and toilets. It is, as previously mentioned, an excellent place on which to put the tail.

The tail consists of two fixed planes: the tailplane, which is the horizontal bit; and the fin, which is vertical; and two moving planes. The horizontal moving plane is called the elevator, and alters the longitudinal attitude of the aircraft about its centre of gravity. More simply, it tips the nose up and down. The vertical moving plane is called the rudder, and acts exactly as the rudder on a boat. There is, however, one catch. If the aircraft banks over (tilts sideways) to an angle exceeding 45 degrees, the functions of the elevators and rudder interchange. Before you reach the critical point, you turn the aircraft by using the rudder, but the steeper the bank you are in, the more the nose will go down, until you 'cross your controls' when you are turning with the elevators and holding the nose up with the rudder. Who can ever forget those moments of instrument flying, when desperately you haul back on the stick, trying to bring the nose up into a climb, while the altimeter unwinds and the speed builds up faster and faster until the moment of horror comes when the instructor hauls back the hood on the link trainer and announces gleefully 'You've crashed!' What you have done, of course, is to unknowingly get past the 45 degree angle, so that hauling back on the stick merely tightened you into a spiral dive.

To finish off the main control surfaces, there are the ailerons. These are on the wings and are so arranged that when one lifts, the other drops. The one that lifts

Flight

makes the wing drop, while the aileron that drops makes its wing lift. This controls the angle of bank, which is very necessary for turning, for were you to turn using the rudder alone, you would only manage a flat skid, which if extreme would ruin the smooth airflow over the lifting surfaces and produce some generally undesirable effects. There are other control surfaces such as slots, flaps, and trimming tabs, but these are rather beyond the scope of this chapter.

The next thing is the cockpit, which is where the pilot lives. For our purposes I am going to discuss single-seat fighter aircraft only. The basic function of the cockpit is to accommodate the pilot plus all the instruments and controls which he needs. Ideally it should be weatherproof, as rain penetrating the instrument panel could have all sorts of unhappy effects. It should afford a modicum of protection for the pilot, both from enemy fire and from the elements. It should also, and this is the catch, afford the best possible view for the least amount of wind resistance, or drag. For the minimum amount of drag, it should be faired as closely as possible to the lines of the fuselage, whereas for the best view the pilot should be sat up as high as possible. Most air forces adopted a compromise, but the early Italian monoplane designs such as the Fiat G 50 and the Macchi 200 opted for maximum visibility, which gave them a distinct hunchbacked profile and, while it afforded excellent visibility, reduced the performance of the aircraft considerably. Pilot protection varied considerably from nation to nation and was improved greatly between 1939 and 1945. The Japanese started out with no protection at all, their theory being that extra weight with its attendant loss of performance was to be avoided at all costs. However, they came to see the error of their ways, and at the end of the war Japanese aircraft carried a fair amount of protection.

Complete pilot protection was, of course, impossible, and had it ever been attempted, the only possible course would have been to cut off the wings and transfer both pilot and aircraft to a tank regiment, as the weight penalty would be so high as to prevent take-off. The general compromise was to stop the light stuff and assorted fragments, and attempt to deflect the heavier bullets. In fact, deflector plates would be a better description than armour. The use of so-called bullet-proof glass became widespread during the conflict. This was a two inch thick toughened glass windscreen, with possibly an extra plate behind the pilot's head so that visibility was not obscured. Most protection was arranged at the front and rear of the pilot, as these were the angles from which he was most likely to get shot up. As the war progressed, other vulnerable parts of the aircraft were also given a degree of protection, notably the fuel tanks.

Fuel tanks are, by their nature, very vulnerable. One small bullet hole and you have petrol swimming around in all sorts of unlikely places, just waiting for a convenient spark. This led to the development of the self-sealing tank which consisted of a coating of crude rubber around the tank which, when in contact with petrol, swelled to seal up the hole from which the petrol came. However, this did nothing for the other hazard. A half-empty petrol tank is, in fact, half full of explosive fumes, awaiting the arrival of one incendiary bullet to turn the whole aircraft into a flaming torch. The counter to this was to fill the tank with an inert gas as it emptied, thereby eliminating the lethal petrol/air mixture.

The next factor is wheels. An aircraft is built to live in the sky, but it has frequently to return to terra-firma for fuel, oil, general maintenance, and pilots' changes of underwear. The simple solution is to fit wheels, but wheels hang down and cause drag. *Ergo*, have a retractable undercarriage. This, unfortunately, involves a weight penalty, and being hydraulic, is something else to get shot up or generally go wrong. The reduction in drag is, however, well worth the added complications. You may think that I am carrying on quite a lot about drag. Consider then, that the drag caused by skin friction lessened by a good coat of wax polish, with particular attention paid to the wing leading edges, can improve the level speed performance of an aircraft by 10 mph or thereabout, and that this represents nearly 300 yards of movement per minute. In a 60 second tail chase you would be nearly 300 yards closer to the enemy, and all due to a good polish.

Finally, we need to consider weapons. All the early British fighters were armed with rifle calibre machine-guns, and most other nations also used them. It was, however, found that unless hits were scored on a vital part, they simple made a small hole. It was not all that unusual for a bomber such as the Heinkel 111 to sustain as many as 200 hits and still get home. Granted it didn't do the morale of the crew much good, and one or two may even have ended up with pneumonia. The Hurricane was also an aircraft which could sustain a great deal of battle damage and still fly; one authority even went so far as to describe it as 'a collection of non-essential parts'. Some aircraft started the war carrying cannon. The arbitrary difference between cannon and machine-guns is that 15 mm calibre and upwards is a cannon, anything less is a machine-gun. Heavy machine-guns of 0.5-inch calibre were also adopted by most combatant nations. The larger the projectile, the bigger the hole it made, and the more chance there was of wrecking something vital. As a generalisation, the larger the projectile, the slower the rate of fire, therefore, less hits. On the other hand, the proportionate damage per hit more than compensated for this. The remaining two factors were the weight of the projectile, which varied widely between nations for weapons of the same calibre, and to a lesser degree the muzzle velocity, or the amount of energy built into a bullet, although this would be variable with the range.

The types of ammunition used were basically ball, explosive, tracer, incendiary and armour piercing. Tracer was to correct the pilot's aim so that he could see where his fire was going. Rather surprisingly, the most effective was not explosive, but armour piercing coupled with incendiary. Explosive shell could go off on hitting a relatively unimportant area of the surface, but armour piercing would plough clean through anything, with any luck releasing all sorts of inflammable fuel, oil, hydraulic fluid or coolant which would be ignited by the incendiary. One final point. The bigger the hole, the less able was self-sealing material to deal with it. Fuel tanks were, therefore, far more vulnerable to the larger size of projectile.

Rockets and bombs were also carried. Air-to-air rocketry was not terribly successful as the rocket was not a very accurate weapon. It was at its best against the close formations of American bombers when it could be fired in salvoes on the shotgun principle. The rocket was better as an air-to-surface weapon, mainly because surface targets tended to be larger and, therefore, easier to hit. Of the bomb, little need be said. Propelled by a gravitational system, they are released in the general vicinity of the target, which they might or might not hit.

Finally, we must consider the other propulsion systems used during the conflict, although these were in service in such small numbers as to be worth only the briefest consideration. The most important is the jet engine, which was flown operationally by the RAF in the Meteor and the Luftwaffe in the Me 262 and the Ar 234. Air is drawn into the front of the engine where it is compressed, then heated which expands it; the general effect being that it comes out of the back faster than it entered the front, thus creating thrust.

The liquid-fuelled rocket engine was the only other type to see combat, in the

German Me 163. All right, I know the pulse-jet powered the V1 but this was a missile rather than an aircraft. Thrust for the rocket was generated by the controlled mixing of two chemical compounds; one highly volatile, the other a catalyst. This type of propulsion has two disadvantages. The first is extremely short duration, measured in minutes only. The second is the lethal nature of the volatile fuel, which killed more pilots in accidental explosions than were ever lost in action. It also had the Hammer Films quality of being able to dissolve the pilot in a few minutes in the event of a leak into the cockpit. In fact, the only time the pilot could consider himself relatively safe was in action.

Chapter 3

Combat flying, aerobatics and tactics

Military aircraft have divers functions which fall into two main categories, defence and offense. The fighter is the prime defensive weapon, the bomber is by its very nature offensive. These categories in turn subdivide quite a lot. The roles of the fighter are a) purely defensive to intercept enemy formations and repel invasions of air space; b) offensive/defensive in escorting one's own bombers, ie, a defensive role in protecting aircraft engaged in a purely offensive activity; and c) the fighter-bomber or ground attack aircraft, which carries out an offensive mission and then has the capability of defending itself on the way home.

The role of the bomber cannot be other than offensive, as regardless of whether it carries bombs, torpedoes or leaflets ('bombphlets', as they were known in 1940), it is deliberately setting out to do a mischief to something, whether it be a factory, a battleship, or the morale of a nation. Dropping leaflets may seem a pretty innocuous pastime, but if someone forgot to cut the string round the bundles and they landed on an unsuspecting enemy head, they could be quite dangerous.

There are two sorts of bombing; strategic and tactical. Strategic bombing is a long-term hammering at the essential industries of the enemy, such as aircraft, steel, oil, etc, hopefully reducing them to the point where his ability to continue the war is seriously prejudiced.

Tactical bombing is behaving like a form of airborne artillery, attempting to give support to one's own ground forces by knocking out communications, supplies, troop concentrations, etc.

The general pattern of bomber aircraft is to fly to the target, plaster it, and come home for tea. To achieve this, one needs a good navigator, an even better bomb-aimer, and no opposition, to say nothing of good weather. The first two conditions might occasionally be met. The third, however, is highly unlikely, so from this we can deduce that, in addition to its active role as a carrier of destruction, the bomber also has a passive role in air warfare. It is a target for any beastliness that the enemy can dream up.

When we consider the attributes of a successful combat aircraft, we must concern ourselves mainly with the fighter. The bomber is generally large, slow, unwieldy (especially when laden with bombs) and generally incapable of looking after itself.

The prime essential of a fighter aircraft is speed. Speed gives the initiative and allows the pilot to choose his own moment to attack as well as permitting him to clear off unmolested if the position is unfavourable or the odds are too great. To paraphrase the old saying, 'He who merely runs away, lives to fight another day'. If a specific aircraft is known to be slower than an adversary, the pilot would be a clot to pick a fight with that adversary on level terms. If, however, he has a significant height advantage, the surplus height can, with the aid of that wonderful invention of Sir Isaac Newton, gravity, be converted into speed, and provided that the extra speed provided by the height advantage is not frittered away in subsequent manoeuvres, the attack can be pressed home with every chance of success. The correct tactic would be dive to build up speed; have a quick squirt, and failing to knock him down on the first pass, either keep the nose down and go home rapidly or, if you feel confident, convert the speed back into height with a zoom climb ready to try again.

The second most important factor is firepower. A great deal of shooting is done in split-second bursts and, while you may pour a lot of lead in the general direction of the target, unless you are an exceptional shot, not very much is going to hit. Two per cent is about the average. This is why the tendency during the war was to introduce heavier and heavier guns, although a balance had to be struck as the larger the gun calibre, the heavier and bulkier it is, thus imposing weight and sometimes drag penalties. The rate of fire also slows more and more as the calibre increases. The effect of the German 30 mm cannon scoring just one hit on a single-seat fighter was awe-inspiring and would almost invariably be enough to bring it down, but the relatively slow rate of fire reduced the chances of a hit being scored at all. The Americans went to the other extreme and hardly used cannon; they were still using the 0.5-inch heavy machine-gun in the Korean conflict. We will deal with guns and gunnery more fully later, mentioning only in passing that whatever the number and size of the armament, the recoil had to be absorbed by the airframe and this could have a distinct slowing down effect.

Rate of climb is the next in order of precedence. Defending aircraft had to be 'up and at 'em' as quickly as possible, in order to intercept effectively. Once action had been joined, the pilot who could climb away most rapidly could gain height and pick his spot to wade in again. Alternatively, if the odds were a bit too long he could always climb away out of trouble and go home. Mind you, aircraft climbing ability could be deceptive, as the best angle and speed of climb varied greatly. For example, the Zero climbed at a very steep angle, giving the impression of an immense rate of climb, but its best climbing speed was very low, and it could be matched by the Spitfire which, although having a much shallower angle of climb, climbed at a much faster speed, thereby cancelling out the steeper angle of the Zero. Remember though that the advantage lay with whoever was doing the chasing in this sort of situation.

The next priority is rate of roll, as this is the pre-requisite of being able to change direction rapidly. If an opponent capable of half rolling in five seconds is on your tail and you can manage it in three, it means a two second advantage in sticking your nose down and heading for the deck, and two seconds at 300 mph is worth nearly 300 yards of distance. The same applies to horizontal turns, although here you would only need 90 degrees of roll instead of 180. Allied to the rate of roll, you would need speed in the dive. Having got an advantage in distance as outlined above, you would need to maintain it. Discretion was the order of the day in air combat; those who didn't possess any generally either didn't get back or learnt some the hard way.

Rate of turn was the final requirement of a fighter aircraft and, surprisingly, the least necessary. It could be very handy provided that the enemy stayed and mixed it, but unless you had one of the other advantages to go with it, it didn't help much.

The correct tactics against a more agile opponent are to come in fast, preferably down sun, a quick squirt then back upstairs like the clappers, circle, and look for an opportunity to repeat the dose. To put it bluntly, you wouldn't give him the chance to whip round on to your tail. This is where the Japanese, with their supremely nimble aircraft, came unstuck against the large and heavy American fighters. They hadn't the speed or climb necessary to force the Americans to fight; the combat was therefore nearly all of the Americans' choosing. Once this lesson had been learnt the Japanese, outnumbered and outflown, simply hadn't an earthly. On the rare occasions when the Yanks decided to get stuck in, they generally came off worst. Consquently the correct tactics were evolved and adhered to.

Now for a brief look at aerobatics. Although very pretty at air displays, they were not really suitable for battle, their main function being to give the pilot confidence. A pilot must function correctly even with his backside pointing skywards, and aerobatics familiarise him with the sensation of the ground being in the wrong place. He may frequently find himself at all sorts of odd angles, and consequently needs to be able to shoot from them in addition to retaining control. As the saying was, 'Forget where the sky is, it's where the enemy is that counts!' The roll in action was definitely *de trop*; its effect was to lose both speed and height and its only possible advantage was to make your opponent laugh and thus spoil his aim. If he was not in a humorous mood you were in dead trouble, the accent being on 'dead'. The loop could be used at a pinch, but only as a species of vertical turn. It could be used to avoid being bounced from astern, but was not as effective as the half roll and reversed turn or even the straight break to port or starboard. The golden rule when being attacked was always face the enemy. Failing this, give him as big a deflection shot as possible. One of the best ways of shaking off a pursuer was to 'stuff everything into the corner', throttle, stick and rudder. The result was ungraceful, and for the pilot quite painful, but as there was no obligation for the pursuer to hurt himself, you would probably get away.

The barrel roll might be used on rare occasions. This manoeuvre, in which the aircraft describes a helical path as though flying around the inside of a cylinder could be of use if, during the general shambles, you found yourself flying alongside an enemy at about the same speed. The greater distance around the helix covered by your aircraft would enable you to slip in behind your adversary *without* having to throttle back, with its resultant speed loss. Of course, this does presuppose that the enemy takes no evasive action, but during the manoeuvre you would be able to keep him in sight all the time and react accordingly. The falling leaf was usually only done by aircraft with the pilot out of action as it is essentially a low-speed manoeuvre. Stall turns could be, and were used in the early days of the war, but lost favour as aircraft became heavier and more unmanageable. As it involved losing speed it was tactically unsound in most circumstances. The sideslip was basically a device for losing excess height on the landing approach and, although its derivative, the flat skid, could get you out of the line of fire pretty quickly on occasion, it did horrible things to the airflow, and could lead to the high speed stall almost invariably accompanied by a spin, and complete loss of control. The spin was an involuntary aerobatic, performed by accident. It might make the enemy think you were finished, but again it might not. It was not recommended as an evasive tactic particularly as some aircraft, such as the F4U Corsair, were notoriously difficult to recover. On recovery you were a sitting duck until you had regained a certain amount of speed.

I realise that I have stressed the overriding necessity of speed, speed and yet more speed. There were, of course, always exceptions, such as the time in 1940 when a Bf 109 and a Spitfire realised that they were on a converging course at an angle of about 30 degrees. They both throttled back and indulged in a slow flying contest which was won by the Spitfire, the outcome of which was one more Luftwaffe gentleman becoming a guest of HM Government. These cases are, however, the exception. Too great a speed was, of course, an embarrassment. If the range at which you could be reasonably expected to hit anything was 300 yards and you were closing with a speed advantage of 200 mph, the time between optimum range and collision point was barely three seconds, which is not really enough to effectively sight, allow deflection, check the turn indicator (to make sure you were not sideslipping), fire a one second burst and still have time to break away. Many pilots who, through inexperience, opened the taps and hurtled down on an unsuspecting enemy, overshot without even managing to bring their guns to bear. The ideal closing speed was always about 50 mph which gave you time to line up your sights, open fire at 300 yards or less, observe your tracers and any hits, adjusting your aim accordingly while continuing to fire, which meant that you could get in a good long burst and still have plenty of time to break off without becoming agitated at the thought of imminent collision. You would also have sufficient margin of speed to evade your infuriated opponent if your shooting had been off target but without an excess margin that would make you overshoot and leave you sitting out in front of him to be clobbered in return. The Germans had excess speed problems in the Me 262. It could hammer along straight and level too fast for anything to catch, but at this sort of speed, it couldn't hit much either. The answer was to attack from above in a shallow dive, pass below the level of the target and then attack on the climb, using the pull-up to 'mush off' excessive speed. Having attacked, they then built up speed to make good their escape.

We must now turn our attention to the limitations of aircraft imposed by altitude. The higher you go, the less air there is. This is both bad news and good news. Let's get the bad news out of the way first. First, there is less air to support you; your stalling speed increases, your manoeuvrability decreases. Then there is less air to be shovelled backwards by the propeller, which means that your speed decreases. This is further aggravated by there being less air to mix with your petrol, so your engine does not produce the same amount of power as at lower altitudes. In the event of a failure in the oxygen supply, the pilot cannot breathe. This reduces his efficiency to say the least. Finally, the more tenuous the air, the lower is the speed of sound and the sooner you hit compressibility problems, commonly known as the sound barrier. On the other hand, the thinner the air the less slowing effect or drag it has. In designing a fighter aircraft compromises had to be made. Engines were designed to give their maximum output at a given altitude, although this could not be done at the expense of high power at ground level for take-off and initial climb. Generally, this was effected by various systems of supercharging, water-methanol and/or nitrous oxide injection which cut in automatically or were manually introduced at suitable heights and speeds. It will be noticed in any study of comparative performance data that the maximum speed is usually attainable in the 18-22,000 feet height range. This is where the power output/reduced drag factor was at its optimum, and performance fell away at both ends of the scale. To be more specific, an aircraft capable of 400 mph in level flight at 20,000 feet would be unlikely to exceed 350 mph at sea level. Aircraft were built specifically for the high and low level roles, but performance loss was always far more marked at the opposite end of the performance spectrum than the middle of the road fighter, which had of necessity to be an all-rounder. Another thing to beware of in published performance figures is the ceiling height. An aircraft might be able to stagger up to 40,000 feet, but its performance would be such that it would be totally incapable of doing anything when it arrived there. Knock 5,000 feet off its ceiling and you probably have something like the maximum height at which it is effective.

The sharp end of a World War 2 fighter; in this case a P-51D Mustang. The drop-tanks, which together with its laminar flow wing, helped to give this aircraft its superlative range, can be clearly seen, as can the three 0.5-inch machine-guns in each wing. Compared with contemporary British and German fighters, the Mustang was relatively under-gunned. The small square hole on the leading edge of the port wing is the camera-gun aperture (*Minyon Prescott*).

The radius of turn also varies with altitude. The lift generated by the wings is about 90 per cent from the upper surface and ten per cent from the lower surface. In a turn the aircraft banks over to prevent it from skidding sideways. The prevention, however, builds up pressure beneath the wings, and the amount of pressure built up is dependent on the density of the atmosphere. At high altitudes the aircraft would have to 'mush' wider in the turn to produce the same amount of pressure; its turning radius would, therefore, be wider, speed for speed, than lower down. A spin-off effect to this pressure build-up is increased drag. The tighter an aircraft turns, the more drag and, therefore, loss of speed. Luftwaffe pilots flying the Ar 234 jet bomber were, in fact, warned to keep their turns gentle, in order to keep their speed above that of the enemy fighters. This was particularly important as the early jet engines suffered from poor throttle response, and speed once lost was only slowly regained.

Chapter 4

Compromise and the needs of wargaming

A friend of mine with considerable flying experience has said to me on more than one occasion 'the only way to get the true feel of air combat is to climb into an aircraft and shoot at somebody who is able to shoot back'. This being so, air wargaming needs to be even more of a compromise than other wargame fields. This chapter is, therefore, being written to explain what compromises have been felt necessary and why, where realism has been departed from, and how to make the necessary calculations if you decide to take to the air in an unusual aircraft such as the Airacuda, or even a 'special' such as a Defiant with fixed forward armament. The main reason for giving the basic calculations is because to cover every type and variant of aircraft ever to see action would be a mammoth task, and most people wouldn't want the information anyway. This way, those who fancy flying something a bit different can calculate their own information and know that it conforms to the data given in this book.

The first step is to decide what scale models you wish to use. The largest range of aircraft which is usable is 1:72 scale; anything larger than this and you will have to hire a football pitch to play on. Even at 1:72 scale you will have to play on the floor, although don't bother about the furniture; just call it cloud cover and fly under or over it. There are a few kits and models manufactured in 1:144 scale, although the range is limited and the actual scales tend to vary. Finally, for those who wish to play on their usual wargames table, there is an extensive range of 1:300 scale model aircraft, which I would recommend. The immediate problem is that, if we stick to a true 1:300 scale, a speed of 400 mph will take you across the average sized room and halfway through the kitchen in ten seconds flat. We therefore need a more realistic distance scale; after much trial and error the most easily usable scale was found to be one millimetre to four yards. This does have the effect of making the aircraft much larger in relation to the distance scale than it should be but on the other hand it does produce a larger target, and correspondingly better results. If you go for absolute realism in results, air warfare is hardly worth playing, as a fairly typical result of 20 fighters milling about would often be two damaged, which is hardly a decisive encounter, and not a very exciting return for an evening's play. This is, then, our first compromise, in the interest of an exciting game.

The next and greatest problem is the third dimension. Accurate dangling from the ceiling proved impossible. Mounting on rods led to wrecked models, broken windows and goosed spectators. It also made cloud cover impossible, the game slow and argumentative, and precision flying out of the question. Therefore, no rods. The next experiment was to play in two dimensions recording height changes.

As a height change during a move obviously affects speed, calculations became complex; the pace slowed. Visually, the two dimensional game played in the horizontal plane was unsatisfactory. The opponents moved, they looked like aeroplanes, they had wings, but they might as well have been armoured cars. Also, no effective cloud cover.

The final answer came during, of all things, a Napoleonic naval game. A third rate was tacking. I suddenly realised that, if I were to substitute gravity in place of wind, and tip the board up vertically, I was watching the equivalent of an aircraft stalling. A trial game was laid on immediately, played in two dimensions; length and height, and it worked perfectly. The sun could be plotted, cloud cover inserted and speed gain or loss in the dive or climb instantly calculated. One side of the board was ground level, with targets and contours roughly sketched in, and altitude measured up from this towards the far edge. World War 2 was very much a matter of dive and climb rather than turning in ever-decreasing circles. The turning of a dogfight had, of course, to be reproduced in the vertical rather than the horizontal plane, but this in itself contained an unexpected bonus, as it tended to produce a speed/height loss which would be the natural result of tight turns. The 'depth' dimension would be marginally introduced with aircraft on a collision course merely passing each other and assuming that they had passed with a lateral space between them, the same assumption being made when a hostile aircraft passes fleetingly across one's nose during the course of a move, the ruling being that, to be a valid target, a hostile aircraft has to be in one's sights at the end of the move.

However, to return to our distance scale of one millimetre to four yards, this makes an eight by six foot board represent an area 5.54 miles long by 21,946 feet high. Not ideal I grant, but usable. If 1:72 scale aircraft are your choice, substitute centimetres for inches and all will be well. This applies right the way through, and a 12 by ten foot floor will scale up to 3.27 miles long and 14,400 feet high. The only disadvantage of the 1:300 scale aircraft is that, while the 1:72nd scale model can be constructed in two halves, each being used depending on which way it is going (see photos), the smaller model has to be seen in plan view but thought of in elevation throughout the game.

Having decided on a model scale and a distance scale, we now need a move time. Air warfare, particularly fighter conflict, is very much a matter of split seconds but, as this is impossible to reproduce in a wargame without slowing it up to an unacceptable degree, I have, after some experiment, adopted a move time of four seconds. This is sufficiently long for a player to commit himself to a manoeuvre, long enough to get in a three-second burst of firing (in practice it is rare to hold the enemy in your sights for longer), long enough to search the sky reasonably well (you must first see the enemy before you fight him, and the sky is a big place), and it has one tremendous advantage. Travelling at 100 mph you are covering 48.89 yards per second, or 50 yards as near as makes no difference. As our distance scale is one millimetre to four yards, the distance covered in four seconds is near enough 200 yards, which at four yards to the mm is 50 mm, or for 1:72 scale, five inches (50 ÷ $\frac{1}{10}$ inch). In other words, halve your speed in miles per hour; this gives you your move distance in millimetres (or tenths of an inch if you prefer the larger scale). Thus all speeds are reckoned in miles per hour rather than so many inches move distance. It gives far more 'feel' to be diving at 416 mph than 208 mm, even if 208 mm is what you have to measure out and move.

Having established this much, it is now time to consider relative aircraft

Compromise and the needs of wargaming

performance, and how to use the mass of performance data, often conflicting, which appears in various publications. Manoeuvrability had better be taken first. Rate of roll which, unfortunately is very important, is just about incalculable for gaming, and must be ignored. It is a complex of factors involving speed, height, wing loading, ratio of aileron area to wing area, factor of mechanical advantage to the control column, wing shape and inertia, and the sheer physical strength of the pilot. There are other factors, but need I go on?

Rate of turn is the other aspect of manoeuvrability. This is comparatively simple, being related to wing loading and speed, although I propose to ignore other things such as combat flaps and altitude on the assumption that two aircraft fighting will both use combat flaps, and altitude differential should cancel out. I have prepared a table (Fig 2) showing the relative wing loadings of different aircraft. These are all based on the normal all-up combat weight but do not include bombs or other external loads. Fighters only are included as we are dealing with maximum rate turns which, if tried in a fully laden bomber, could have catastrophic results. Whilst it should always be remembered that some bombers, such as the Sparviero, were fully aerobatic, the usual bomber defence tactic was to bunch tightly together for mutual protection like maiden aunts in a soccer crowd. Different criteria apply, with which we will deal later.

Our first need is to equip ourselves with a turning device, which should be as Fig 1 (which is not to scale), and scribed on to transparent Perspex with Letraset lettering. The inner circle '2' should be two centimetres radius (two inches for 1:72 scale) and each successive circle should be one centimetre (or inch) wider than its predecessor.

Our next task is to sort out satisfactorily the turning radii for different aircraft. A glance at Fig 2 shows the Gladiator II at approximately 15 lb per square foot wing loading and the Me 262 at 60 lb per square foot. These then are the parameters within which we must work. A brief aside at this juncture; four Marks of Spitfire are shown, and it can be seen that as the war progressed and aircraft got faster, they also got heavier; wing loadings, and therefore rate of turn, increased in proportion. This was true of all developing aircraft types of all nations and is confirmation of the fact that speed is more important than manoeuvrability.

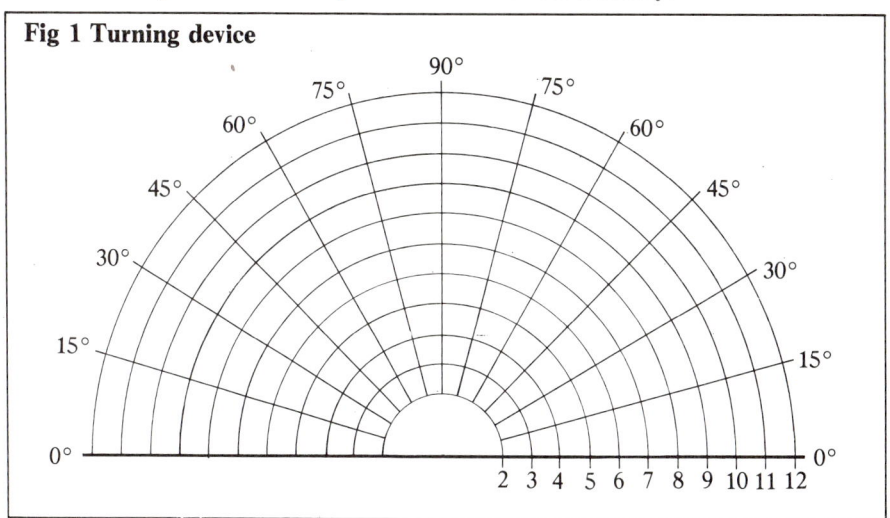

Fig 1 Turning device

Fig 2 Wing loadings (at normal take-off weight)

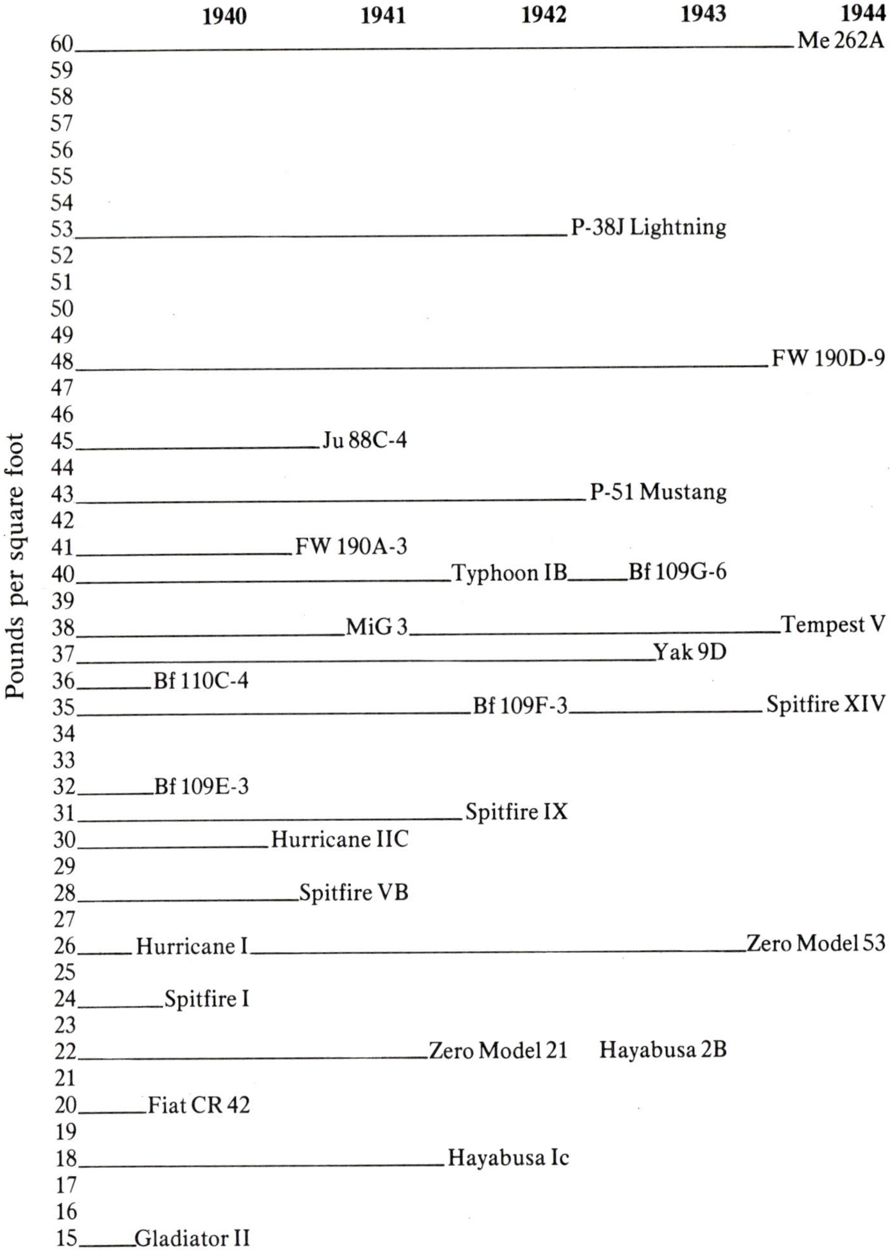

```
                    1940         1941         1942         1943         1944
         60_____Me 262A
         59
         58
         57
         56
         55
         54
         53_____P-38J Lightning
         52
         51
         50
         49
         48_____FW 190D-9
         47
         46
Pounds   45_____Ju 88C-4
per      44
square   43_____P-51 Mustang
foot     42
         41_____FW 190A-3
         40_____Typhoon IB_____Bf 109G-6
         39
         38_____MiG 3_____Tempest V
         37_____Yak 9D
         36_____Bf 110C-4
         35_____Bf 109F-3_____Spitfire XIV
         34
         33
         32_____Bf 109E-3
         31_____Spitfire IX
         30_____Hurricane IIC
         29
         28_____Spitfire VB
         27
         26_____Hurricane I_____Zero Model 53
         25
         24_____Spitfire I
         23
         22_____Zero Model 21   Hayabusa 2B
         21
         20_____Fiat CR 42
         19
         18_____Hayabusa Ic
         17
         16
         15____Gladiator II
```

We must now consider speed as a function of turning rate. When in level flight, the pilot and aircraft are subject to the normal gravitational effect, or 'g'. When turning at speed, the 'g' builds up, and in extreme cases, both pilot and aircraft could experience their apparent weight being six or seven times the normal, or 6 or 7 g. Most of you will have experienced the sensation at the fairground while being

Compromise and the needs of wargaming 27

flung around on the Whiplash or something similar. As we have seen previously, an aircraft's lift is generated by drag. In a turn, the drag increases speed for speed, proportionately to the amount of 'g' being induced. As drag in level flight increases in proportion to the square of the speed, we now have enough information to work on. A glance at Fig 3 shows how drag increases as the speed rises. If we make an arbitrary assumption that the average aircraft is able to turn tightest at 200 mph, then 200 mph compared with 600 mph gives a turning rate differential of 40:360, or 1:9. Simplified, this means that an aircraft travelling at 600 mph would use nine times the radius of turn that would be needed at 200 mph. Therefore, if the radius of turn at 200 mph is 2 cm, the radius at 600 mph would need to be 18 cm, with seven intermediate stages. This is not entirely correct for aerodynamic reasons; it is also too unwieldy for our purposes; we must consequently make some arbitrary adjustments. After all, compromise is the very essence of the wargame, unless your name is Adolf and you are in a position to start a real war.

The first compromise is to produce a smaller turning circle which is the one

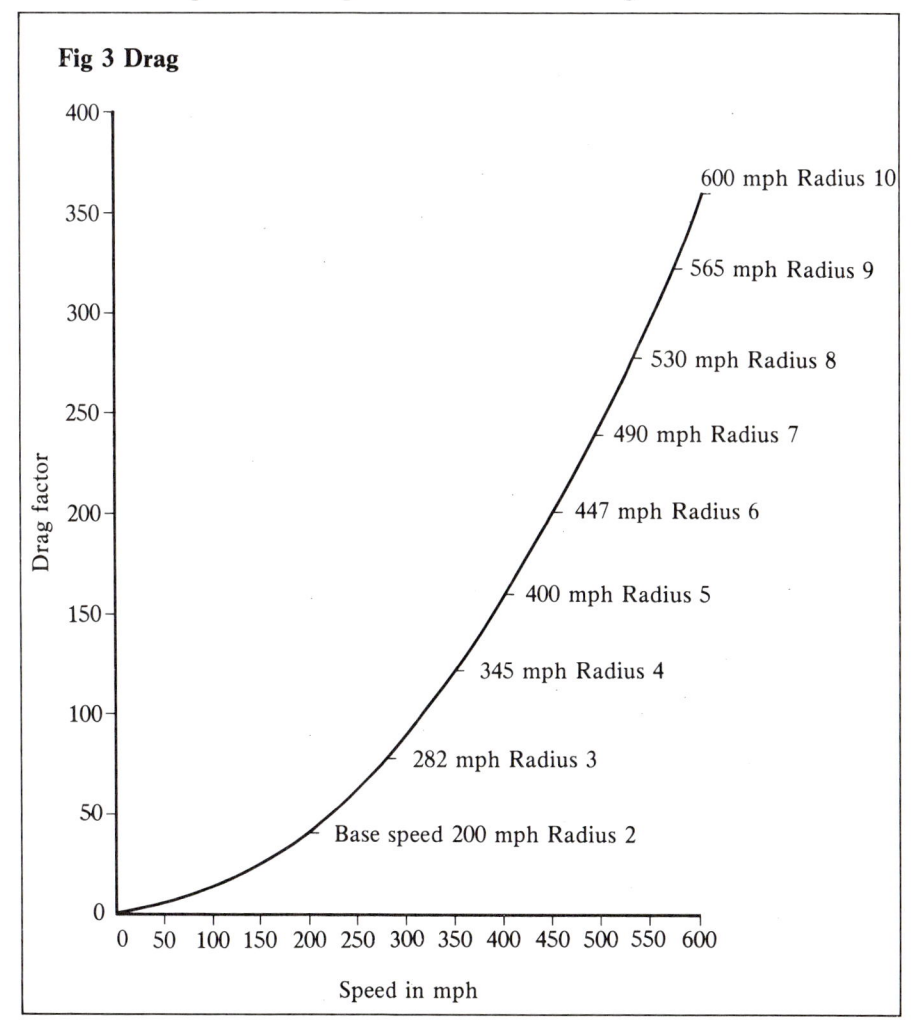

Fig 3 Drag

Drag factor (y-axis: 0 to 400)
Speed in mph (x-axis: 0 to 600)

- 600 mph Radius 10
- 565 mph Radius 9
- 530 mph Radius 8
- 490 mph Radius 7
- 447 mph Radius 6
- 400 mph Radius 5
- 345 mph Radius 4
- 282 mph Radius 3
- Base speed 200 mph Radius 2

already described, and to make the calculated performance data fit. Working exactly to Fig 3, an aircraft capable of attaining 600 mph in a dive would need nine turning radii; the base radius, widening by one ring at speeds of 282, 345, 400, 447, 490, 530 and 565 mph. Without getting too far adrift, we could adopt speed ranges

Fig 4 Turning abilities

| WL (Pounds) | Stall speed | \multicolumn{8}{c}{Radii—centimetres or inches} ||||||||
		2	3	4	5	6	7	8	9
15	56	274	384	484	564				
16	57	259	373	475	557				
17	58	244	362	466	550				
18	59	229	351	457	543				
19	60	214	340	448	536				
20	61	199	329	439	529				
21	62	184	316	428	520	592			
22	63	169	303	417	511	585			
23	64	154	290	406	502	578			
24	65	139	277	395	493	571			
25	66	—	264	384	484	564			
26	67	—	251	373	475	557			
27	68	—	238	362	466	550			
28	69	—	225	351	457	543			
29	70	—	212	340	448	536			
30	71	—	199	329	439	529			
31	72	—	184	316	428	520	592		
32	73	—	169	303	417	511	585		
33	74	—	154	290	406	502	578		
34	75	—	—	277	395	493	571		
35	76	—	—	264	384	484	564		
36	76	—	—	251	373	475	557		
37	77	—	—	238	362	466	550		
38	78	—	—	225	351	457	543		
39	79	—	—	212	340	448	536		
40	80	—	—	199	329	439	529		
41	81	—	—	184	316	428	520		
42	82	—	—	169	303	417	511		
43	83	—	—	154	290	406	502		
44	84	—	—	139	277	395	493		
45	85	—	—	—	264	384	484	564	
46	86	—	—	—	251	373	475	557	
47	87	—	—	—	238	362	466	550	
48	88	—	—	—	225	351	457	543	
49	89	—	—	—	212	340	448	536	
50	90	—	—	—	199	329	439	529	
51	91	—	—	—	184	316	428	520	592
52	92	—	—	—	169	303	417	511	585
53	93	—	—	—	—	290	406	502	578
54	94	—	—	—	—	277	395	493	571
55	95	—	—	—	—	264	384	484	564
56	96	—	—	—	—	251	373	475	557
57	97	—	—	—	—	238	362	466	550
58	98	—	—	—	—	225	351	457	543
59	99	—	—	—	—	212	340	448	536
60	100	—	—	—	—	199	329	439	529

Compromise and the needs of wargaming 29

of: stall to 199/200—329/330—439/440—529, and 530 upwards. This will give even the fastest aircraft a maximum of five turn rates, and the slowest ones as few as two, which is much more easy to manage.

The final factor in assessing rates of turn is to establish the maximum turn rate at low speed for each type of aircraft, which is done by returning to Fig 1 to compare wing loadings. Although the aircraft listed are not particularly comprehensive, they are fairly typical, and it can be seen that the main spread of wing loadings comes between 25 and 50 lb per square foot. If we can once again be arbitrary, and say that 1 lb per square foot wing loading equals one millimetre (or tenth of an inch) radius of turn at a speed of 199 mph, this gives us a working basis. This means that an aircraft with a wing loading of 30 lb per square foot flying at 199 mph will turn on the 3 cm radius, while one with a 40 lb per square foot wing loading at the same speed turns on the 4 cm radius. This is fine for aircraft whose wing loadings are in exact multiples of ten pounds, but we must interpolate for the 'in between' wing loadings. This is done by adjusting the speeds at which the turn rate changes within the limits already laid down, so that each aircraft type turns according to its wing loading, full details of which are given in Fig 4. Look down the wing loading (WL) which is given in pounds, and read off the turning ability of that aircraft in the line against it. For example, the Spitfire Mk I has a wing loading of 24 lb. Against the 24 lb line, we find that the stalling speed is 65 mph, and the rates of turn are as follows: stalling speed to 139 mph is on the 2 cm radius, 140 mph to 277 mph is on the 3 cm radius, 278 to 395 mph is on the 4 cm radius, 396 to 493 mph is on the 5 cm radius, 494 to 571 mph is on the 6 cm radius, and speeds above this are on the 7 cm radius. The Spitfire I, however, has a terminal velocity of 444 mph so the final two radii will not count for this particular aircraft.

One final thing you will need is a table of turning radii distances to speed calculation during the game, which is given in Fig 5. For example, if your speed is 342 mph, your move distance will be 171 mm and you wish to turn 30 degrees upwards, then straighten your climb at this angle, and your aircraft has a wing loading of 32 lb per square foot, meaning that its tightest radius of turn at this speed is on 5, 30 degrees of turn on this radius reads off as 26 mm, thus leaving a further 145 mm of movement in a straight line.

Fig 5 Turning radius distances

Radius	5°	10°	15°	30°	45°	60°	75°	90°	Radius	5°	10°	15°	30°	45°	60°	75°	90°
2	2	4	6	11	16	21	27	32	8	7	14	21	42	63	84	105	126
3	3	6	8	16	24	32	40	47	9	8	16	24	47	71	94	118	141
4	4	8	11	21	31	42	53	63	10	9	18	26	52	79	104	130	157
5	5	9	13	26	39	52	65	79	11	10	20	29	58	86	116	145	173
6	5	10	16	31	47	62	78	94	12	11	22	32	63	94	126	158	189
7	6	12	19	37	55	74	93	110									

The radii are given in inches (centimetres) and the distances per degree of turn are in tenths of an inch (millimetres).

The information so far applies only to fighters, fighter-bombers, dive-bombers, and single-engined torpedo bombers. Level bombers and torpedo bombers developed from level bombers are a different case as, whilst a few were fully aerobatic, most were not stressed for violent manoeuvres, especially with a load on board. Turn rates on level bombers can, therefore, be simplified thus: 10 cm for twin-engined aircraft, 11 cm for trimotors, and 12 cm for four-engined and larger aircraft. For game purposes fighters can generally change direction in the vertical plane with a half-loop, but it would be unrealistic to cut out a horizontal turn altogether, although it must of necessity be a full 180 degrees. A half-loop would be impossible for a laden bomber anyway. The method of carrying this out is as follows. The pilot announces that he is going to carry out a flat turn. A fighter does this in two moves, each of half his normal move distance. If diving or climbing, the same angle of dive or climb is maintained as shown in the sketch.

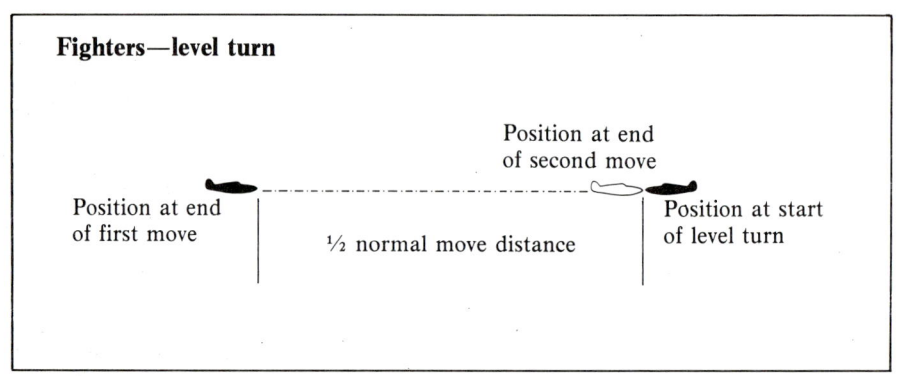

Fighters—level turn

Position at end of second move

Position at end of first move

½ normal move distance

Position at start of level turn

A bomber, however, takes three moves to carry out this manoeuvre. Having announced his intention of flying a reciprocal course, he then makes a straight move of two thirds of his normal move distance. In the next move the aircraft reverses direction but does not change position, merely pivoting about its nose. In the following move it once more makes a straight move of two thirds the normal distance, and at the end of this is deemed to have turned through 180 degrees at the original height, see diagram.

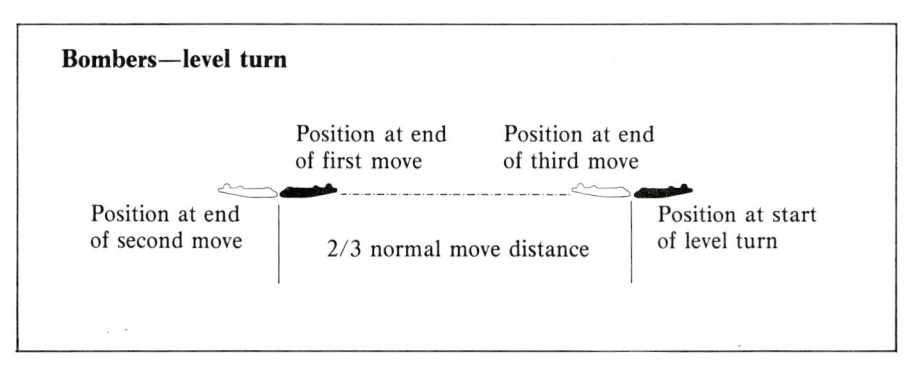

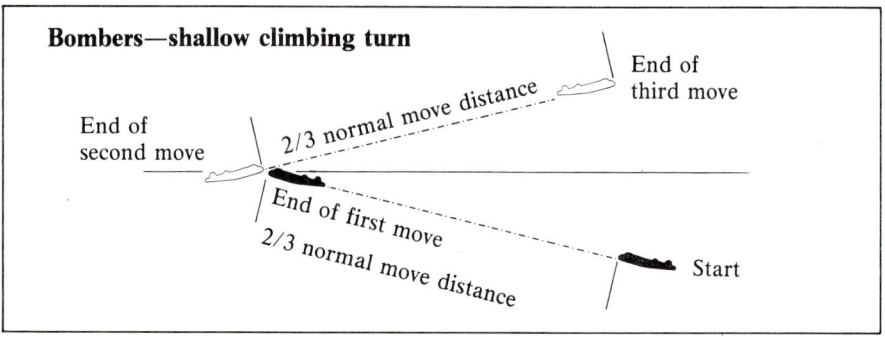

Bombers—level turn

Position at end of first move

Position at end of third move

Position at end of second move

2/3 normal move distance

Position at start of level turn

Bombers—shallow climbing turn

End of second move

2/3 normal move distance

End of third move

End of first move

2/3 normal move distance

Start

Compromise and the needs of wargaming 31

Having now covered manoeuvrability we must next examine accleration. This is rather a difficult thing to assess, as it varies according to both height and speed even with the same aircraft. In addition, there are anomalies which may or may not be subjective but are difficult to account for. To quote but one example, a Finnish report on a captured Russian LaGG 3 states that acceleration was poor. The Japanese, however, were of the opinion that its acceleration was one of its outstanding qualities. It appears, therefore, that we must cover the subject of acceleration in a rather cavalier manner, otherwise it will be easy to bog down in a welter of technicalities. For wargame purposes we need something cheap and cheerful, and preferably enhanced sufficiently to liven the game up. After much trial and error, I have come to the conclusion that ten per cent of the maximum straight and level speed of the aircraft, rounded off to the nearest mile per hour, is most satisfactory for our purposes, and if we limit ourselves to three throttle settings of open/cruise/shut, this will give sufficient variation for an interesting game. The open throttle setting will give acceleration as stated at all speeds up to the maximum level speed. In a dive this can be exceeded but gravity tends to take over as the main propellant until, as the aircraft nears its terminal velocity, the slipstream is forcing the engine to over-rev, and even with the throttle wide open the engine has a braking effect on the aircraft. To give a simple approximation of this without involving ourselves in some very complicated mathematics, the easy thing to do is to assume a positive deceleration of 2½ per cent at speeds above the maximum level speed even with the throttle open, any speed increase above maximum level speed being solely due to the force of gravity, hereafter referred to as Sir Isaac. The cruise setting gives nil acceleration up to maximum level speed, and a positive deceleration of five per cent above this. The shut throttle setting gives a positive deceleration of five per cent at speeds up to maximum level, and ten per cent above this.

Let us take as a specific example an aircraft with a maximum level speed of 360 mph and a terminal velocity of 450 mph.

	Speed range	**Acceleration/deceleration**
Throttle open	Up to 360 mph	+36 mph
	361-450 mph	−9 mph
Throttle cruise	Up to 360 mph	Nil
	361-450 mph	−18 mph
Throttle shut	Up to 360 mph	−18 mph
	361-450 mph	−36 mph

Just one point here. When in doubt, round fractions upwards. If this is made standard practice, no arguments can ensue.

The next section is devoted to Sir Isaac Newton who, as you may know, had an apple dropped on his head by a Stuka. After the usual profanities he postulated a universal law to the effect that what goes up must come down. This naturally applies to aircraft and thus we must give it full consideration. An aircraft in flight is constantly opposed to Sir Isaac, and only maintains itself in the air while it has more power than he has. The effect of gravity has many faces. It affects the rate of climb (think of a car going up a steep hill, it's the same effect), accelerates a dive without really trying (car going down a steep hill) and governs the height which the aircraft can reach because, as engine power falls away in rarefied air, the power output reduces to the stage where it can do no more than balance the effect of Sir Isaac. *Ergo*, no more climb, no more height. What we are really concerned with here, however, is the effect it has on climb and dive capabilities.

The rate of climb is usually fairly easily obtained from performance details published in many books. However, this is frequently given in two different forms. One is initial climb rate, the other gives a time to a specified altitude. Climb rates tend to vary at different altitudes but, as we are dealing with aircraft fighting within a similar height band, as long as our information comes from roughly comparable data, the comparison is valid for our purposes. If we then always use the time to a specified altitude, usually between 15 and 20 thousand feet, we can arrive at conclusions which are reasonably fair. A glance at Fig 6 gives us an idea of the climbing abilities of different aircraft in feet per minute of altitude gained. The aircraft shown are a mixed bag covering all nationalities but are a fair representation of World War 2 fighters. As we are considering maximum climb rates, the throttle will, of course, be fully open. The climb rate then becomes a contest between the power available and Sir Isaac. Aircraft tend to climb best at different angles and speeds according to type. For example, comparative tests between a Zero and a Seafire established that the rate of climb was similar, but the best climbing speed of the Zero was about 125 mph while the best climbing speed of the Seafire was about 165 mph. This means that both aircraft attained the same height in approximately the same time, but the angle of climb for the Zero would be much steeper than that of the Seafire, as the track of the slower moving aircraft would necessarily be shorter, see sketch, in which the tracks are to scale although the angles are exaggerated.

While the precise angle of climb of these two aircraft can be calculated, the necessary information is not readily available for most others. We therefore need a rule of thumb which can be applied to published climb performance data, and does

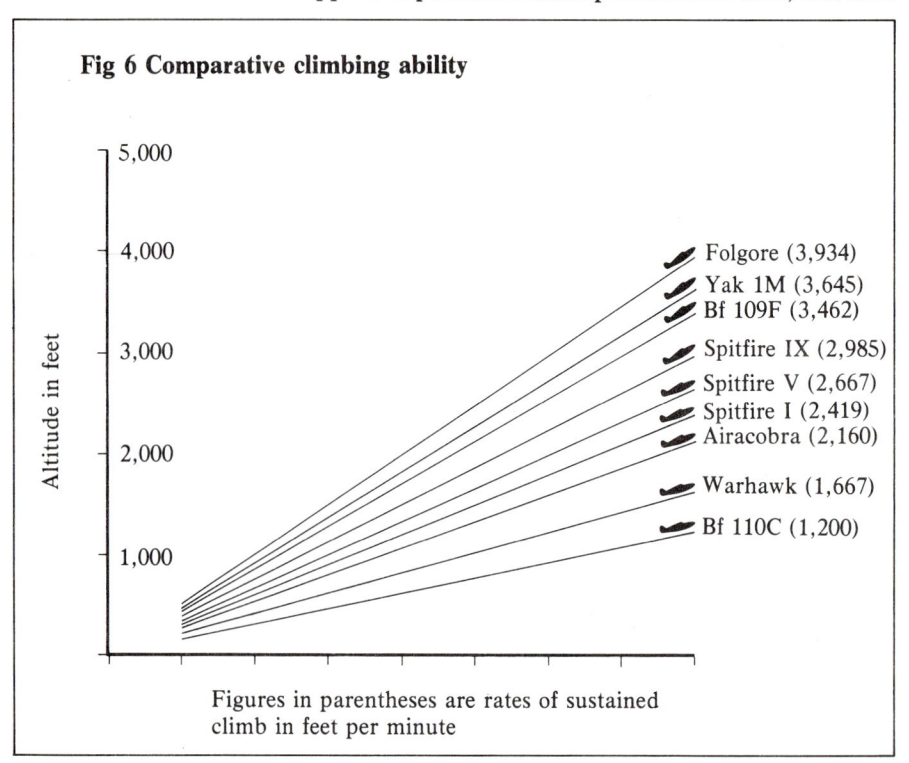

Fig 6 Comparative climbing ability

Folgore (3,934)
Yak 1M (3,645)
Bf 109F (3,462)
Spitfire IX (2,985)
Spitfire V (2,667)
Spitfire I (2,419)
Airacobra (2,160)
Warhawk (1,667)
Bf 110C (1,200)

Figures in parentheses are rates of sustained climb in feet per minute

Compromise and the needs of wargaming 33

Climb rate—Seafire versus Zero

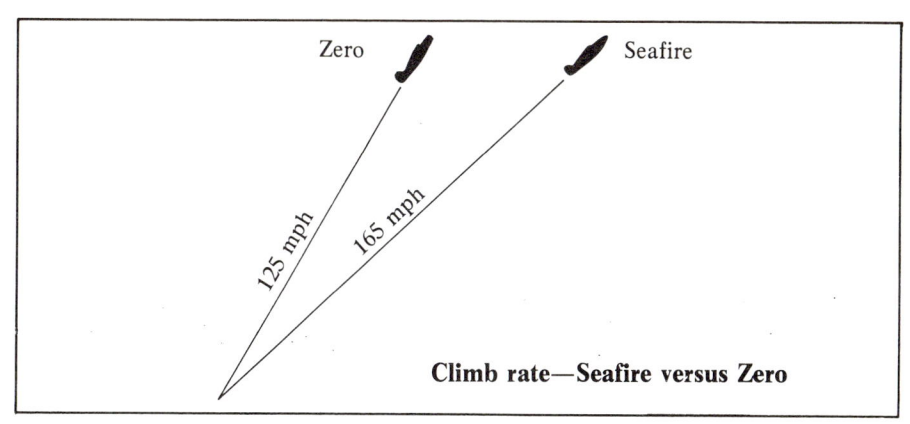

not need Dr Barnes Wallis to calculate. We have already established an acceleration formula, which I am only too willing to admit is greatly exaggerated. If we also enhance the climb and dive formulae, this will have the effect of making things happen faster during the game, and will compensate to some degree for the four-second move time and consequent loss of split-second reactions, also bearing in mind the absolutely gigantic size of the models in relation to the true ground scale. What I propose then is that the acceleration be made to cancel out at a preset climbing angle, in other words an aircraft climbing at full throttle at the required angle will neither gain nor lose speed, but continue to gain height at the same speed at which it entered the climb. A steeper climb and it will lose speed, a shallower climb and it will accelerate slightly.

The way to do this is to plot a chart on graph paper as Fig 6, with an eight-inch base line, and height represented vertically at one thousand feet to the inch. On the altitude line, mark the rate of climb in feet per minute for the aircraft under consideration, then draw a line between this mark and the left-hand end of the base line. This will give a suitable wargame type angle of climb to the aircraft under consideration. Fig 6 contains nine examples and you will see that they differ quite appreciably, certainly enough to highlight the actual performance differences.

Now for the actual calculation. The Folgore at one end of the scale will provide a useful start. Its maximum speed is 370 mph, thus its acceleration will at ten per cent be 37 mph per move. Divide the acceleration factor by the rate of climb in thousands of feet, in this case 3.934, and this, rounded off to the nearest whole number, will give you a factor for the penalising effect of gravity, 9 mph for each centimetre of height gained. We have now established that the Folgore will climb at a steady 4 cm per move at full throttle without losing speed. Let us have a few worked examples.

Example 1 Macchi C202 Folgore at its optimum climb angle, with a calculated speed at the end of the previous move of 216 mph.

216 mph gives a move distance of 108 mm.
(1) Height recorded at the end of the previous move 62 centimetres.
(2) Height recorded at the end of the present move 66 centimetres.
Recalculate speed at end of move as follows:
Speed at start of move 216 mph.
Throttle open, add acceleration factor of 37 mph = 253 mph.
Height at end of move 66 cm, height at beginning of the move = 62 cm, thus height gained during the move = 4 cm.

Climb—example 1

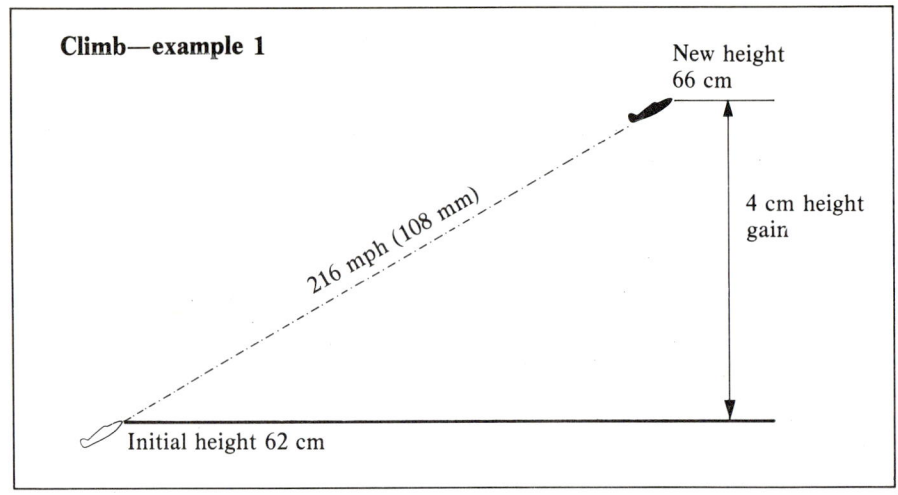

New height 66 cm
4 cm height gain
216 mph (108 mm)
Initial height 62 cm

Speed loss per centimetre of height gained is 9 mph, therefore total speed loss during the move is 4 × 9 = 36 mph.

253 mph minus 36 mph = 217 mph.

Easy, isn't it? The 1 mph discrepancy is, of course, due to rounding off in whole numbers. You just can't win them all.

Example 2 Our old friend Folgore doing a steep climb at 60 degrees, speed 216 mph as before.

Move distance 108 mm.

(1) Height at start 50 cm.

(2) Height at end 59 cm (always round off to nearest, any discrepancy will be taken care of in the next move).

Recalculate speed at end of move as previously.

Speed at start of move 216 mph.

Throttle open, add acceleration factor 37 mph = 253 mph.

Height gained during move 59 minus 50 = 9 cm.

Speed loss during the move = 9 cm × 9 mph = 81 mph, therefore speed at the end of the move = 253 minus 81 = 172 mph, a total speed loss of 44 mph during the move.

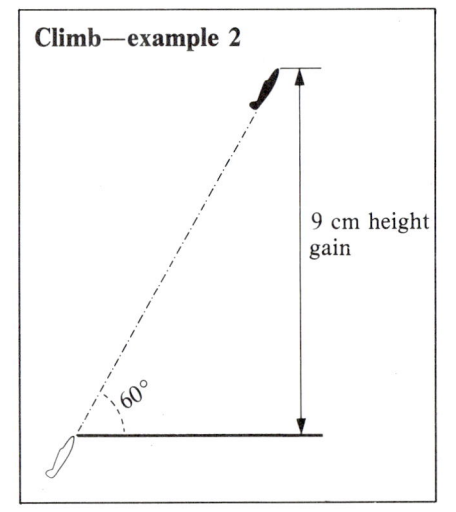

Climb—example 2

9 cm height gain
60°

Having considered Sir Isaac's effect on climb, we must now accord him the honour of a section all to himself. What goes up must come down is Sir Isaac's motto, and he makes darned sure that it does. Running downhill on a bike it is not unusual to pick up speed even without pedalling. The same is true of flying. In a dive one tends to accelerate whether one wants to or not.

An aircraft which is good in the climb is frequently good in the dive. However, an aircaft with the climb rate of a tired brick is frequently excellent in the dive due to its sheer weight, and it is on weight alone that we evaluate the diving performance. For example, the Hurricane II could outclimb the Ju 88C-4 quite comfortably, but

Compromise and the needs of wargaming

in a dive it was in real trouble. Although much the same on the level, the 3½-ton Hurricane was no match for the 12-ton Ju 88 when going downhill. Throttle acceleration will be as before, but acceleration due to height loss may be usefully calculated at 2 mph per ton weight of aircraft per centimetre (or inch) of height lost divided by the number of engines doubled for single engined aircraft; the number of engines squared for multi-engined aircraft. If we take a Hurricane at full throttle, commencing speed 250 mph diving at 60 degrees to escape a Ju 88C-4 at the same commencing speed diving after it like the hounds of hell, the results are as follows, with the height loss of 11 cm. The Hurricane has throttle open and accelerates at 32 mph. It loses 11 cm of height at 3 mph acceleration per centimetre = 33 mph, ie, a total speed gain of 65 mph. The Ju 88, however, gains 30 mph with the throttles open, but its 11 cm of height loss are worth 6 mph per centimetre which equals 66 mph, plus 30 mph = 96 mph total speed gain, an advantage of 31 mph over the Hurricane.

To compensate for the rather long move time when compared with the reaction time of the average fighter pilot we are putting together 'time sequences' when things happen much faster than in real life, and in fact fast enough to set the wargamer, whose skin is not at stake, some very real problems. For instance, the gamer would be unlikely to overshoot his target if the dive and climb rates were slower; nor would he be likely to be left floundering by a rapid change of direction on the part of his opponent. This, I think you will agree, would be unrealistic. Also it is some compensation for the loss of a playing dimension to have these exaggerated performance figures.

A few other aspects of flight need to be considered, although not in so much detail, and perhaps a few things should be defined. To take the various speeds; we first have stalling speed. This is the speed at which the airflow over the lifting surfaces ceases to provide the lift to keep the aircraft airborne, the result being that it falls out of the sky. This must, of course, be reproduced on the table. An aircraft which, at the end of its move, has its speed calculated at on or below the stalling speed has stalled. Its recorded speed falls to zero and for its next move, it pivots about its nose until it is pointing vertically downwards. At this point it must check to see if it spins. (Spinning will be dealt with later on.) At the end of the first stall move the pilot *must* open the throttle wide; the aircraft will then commence a vertical dive until flying speed, ie, any speed higher than stalling speed, is reached, when the aircraft will be back under control and able to fly normally. Stalling is not recommended at low level.

Stalling speed is frequently available from published data but where it is not, the following is a rough guide, not taking things such as flaps, etc, into account. The Spitfire I, with a wing-loading of 24 lb per square foot, stalled at about 65 mph, while at the other end of the performance envelope, the Me 262 with a wing loading of 60 lb per square foot stalled at about 100 mph. Fit these two extremes on to graph paper, draw a line, and read off stalling speeds in terms of wing loading. This is how I have arrived at the stalling speeds given in Fig 7.

An aircraft stalling even in level flight will often drop a wing and spin. An aircraft stalling in combat is unlikely to be in level flight and the probability of a spin is very high in consequence. If we then assume that the probability of a spin occuring when an aircraft stalls is 70 per cent, then the possibility of recovery at the end of the first move of spinning after flying speed has been regained can be arbitrarily assessed as 30 per cent, reducing by 5 per cent for each further move of spin until at the end of the sixth consecutive move spent spinning (24 seconds of

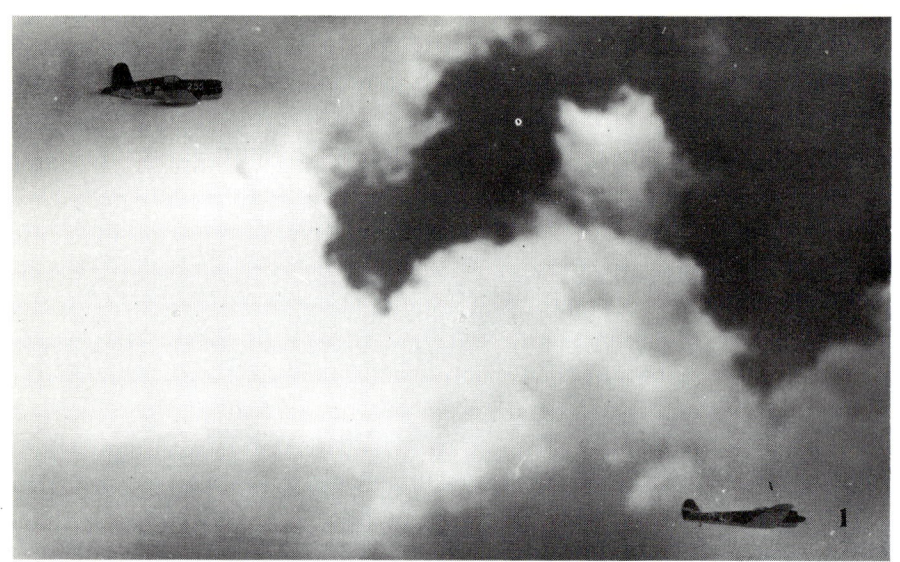

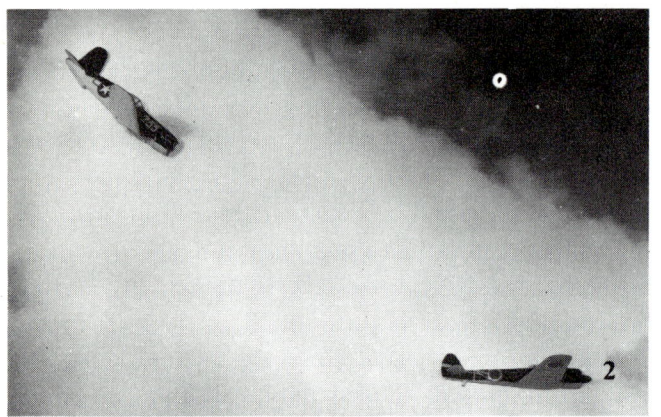

Be careful who you take on. **1** Hank, one of the famous 'Flying Rubbernecks', manoeuvres in the perfect position for a bounce. He has height and position; and his F4U Corsair is faster and more heavily armed than the Japanese Hayabusa he is stalking, flown by 'Kamikaze' Smith. **2** Hank closes the throttle and moves gently down on 'Kamikaze's' tail. He is holding his fire to get a point-blank, no deflection shot. 'Kamikaze' has seen him though, and is biding his time. **3** 'Kamikaze' slams his throttle open and breaks hard upwards, leaving Hank with no target. His Corsair cannot match the Hayabusa in a turn.

Compromise and the needs of wargaming 37

4 Hank instead of breaking off, opens his throttle and tries to follow 'Kamikaze' round. No chance. The light and agile Hayabusa is clear away. **5** Hank is still pulling over the top, regretting that he didn't attack at full throttle, as 'Kamikaze' is already round on his tail. He lets fly, but fortunately for Hank his two light machine-guns don't do much damage.

6 Hank comes over the top and half-rolls upright. He cannot turn out of the line of fire as the Hayabusa is so much more manoeuvrable. 'Kamikaze' scores two more hits, in the wings and fuselage, which do little damage. **7** Hank, much chastened and perforated does the only possible thing. He sticks his nose down and dives away, his heavy Corsair easily outdistancing the Hayabusa. Had the Japanese aircraft carried cannon he would have been shot to pieces (*except as noted otherwise, all photos by Brian Monaghan*).

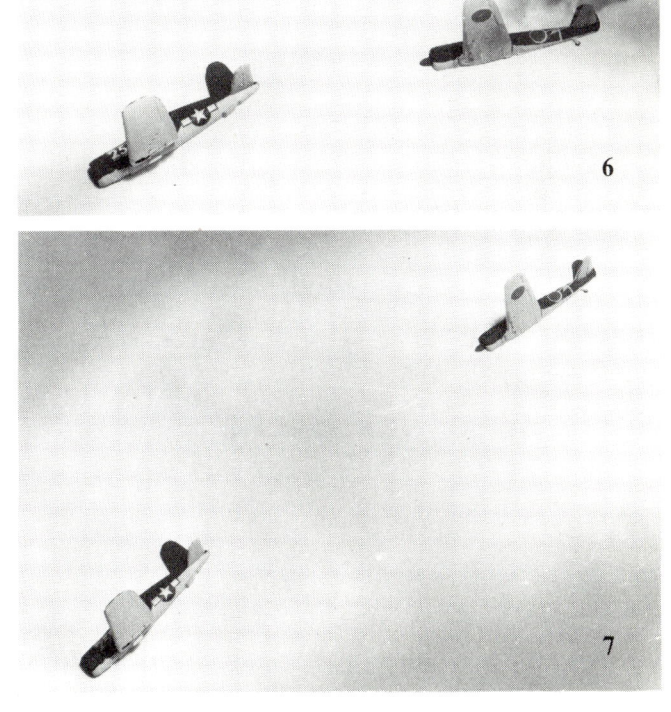

Fig 7 Stalling speeds

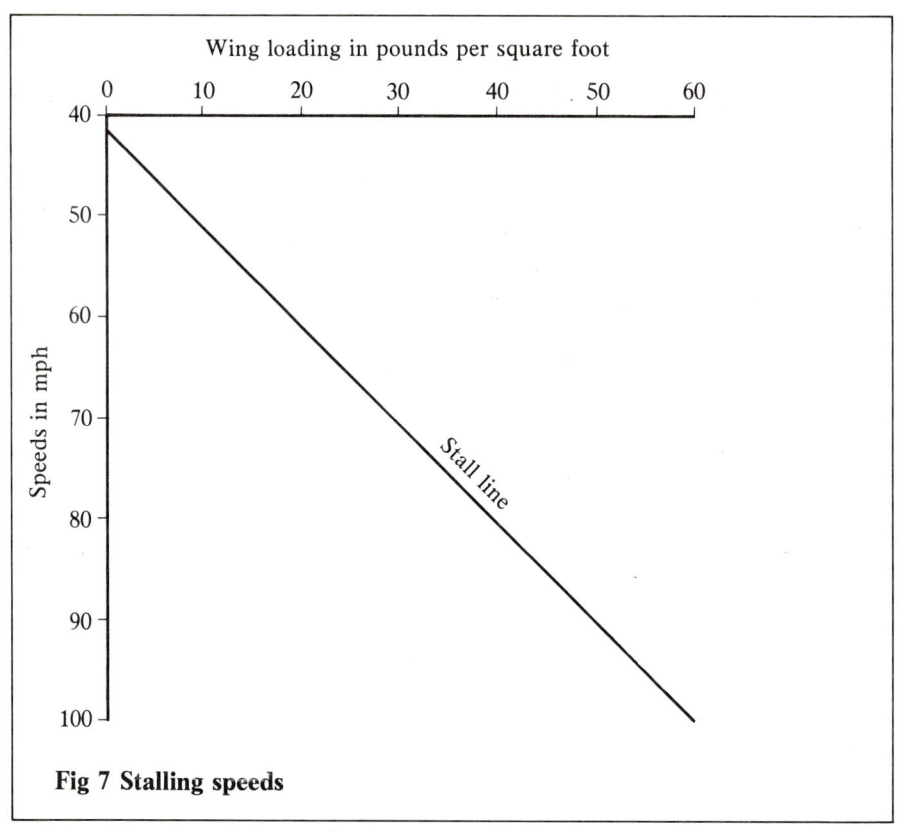

real time) the aircraft is completely uncontrollable. It was nothing unusual to lose 5,000 feet in a spin, and as speed builds up it becomes increasingly more violent and difficult to control. Different aircraft had differing spin characteristics, but to attempt to reproduce them in a game is taking realism too far.

Whilst we are in the area of loss of control, there is also a thing called a high speed stall. It is vicious, frequently irrecoverable, and comparatively rare, so I have made no attempt to reproduce it in the rules.

Cruising speed divides into two distinct categories. Economical cruising, which is more applicable to a campaign or map game, and high cruising, which is the speed most likely to be used in the combat area, before the fighting actually starts. Both cruising figures are occasionally quoted in aircraft reference books, but for a reasonable approximation of high cruising, deduct 50 mph from the maximum speed in level flight, unless this is below 300 mph when five sixths of the maximum level speed can be assumed.

Maximum level speed is the fastest speed attainable in level flight at the optimum altitude, frequently the 18 to 20 thousand feet area, and dropping considerably at sea level. However, as two aircraft trying to clobber each other will be in the same height range, the altitude/speed variation doesn't matter as much as you might think. Maximum level speed is one thing you will have no difficulty in obtaining from reference books. The only problem is that it varies from book to book. As individual aircraft of the same type varied slightly in performance, perhaps you should use the performance data given later.

Compromise and the needs of wargaming 39

Terminal velocity is the maximum speed attainable by the aircraft in a steep dive. It occurs when the drag builds up sufficiently to counteract the combined efforts of both the engine and Sir Isaac. This information is sometimes available in books, but more often it is not. The rule of thumb in this case is maximum level speed plus 25 per cent so if, for example, MLS = 360, TV = 360 + 25 per cent (90) = 450 mph. From 1943 onwards, as aircraft got faster, the so-called sound barrier started showing itself in the form of buffeting and general lack of control, caused by the compressibility effect. A direct result of this was that some aircraft, particularly the German jets, were given a limiting airspeed which was not to be exceeded.

While we are on the subject of extremely high speeds, it should be noted that various forms of emergency power were used for short periods, the *Luftwaffe* for example using water-methanol and nitrous oxide. As most published performance data give speeds attainable by these means, I have considered it only fair to use these maximum speeds as the norm.

External loads were often carried, such as bombs, rockets, or fuel tanks. These quite naturally had an adverse effect on performance, but the effects of different combinations of loads would be extremely complex to calculate as, whilst the additional weight is easy to assess, the drag factor is not. As a general rule, we can assume that external loads have the following effects on performance.

Turning circle—in accordance with the new wing loading.
Maximum level speed—minus 50 mph.
Acceleration—revise in accordance with the new maximum level speed.
Rate of climb—minus 20 per cent.
Dive—unchanged unless the extra weight exceeds half a ton—recalculate the dive formula.
Stalling speed—in accordance with the new wing loading.

Nasty I know, but if you will hang the kitchen sink on your aircraft what do you expect?

A final word to close the chapter on how to go about flying a 'special'. Let us, for example, take our old friend the Defiant. Apart from the actual, we can use him in two different ways, the first being with fixed forward armament of 4 × .303-inch Brownings. This is a quite simple weight adjustment and involves adding the weight of the four-gun installation plus ammunition, approximately 200 lb, to the normal loaded weight of 8,318 lb, an increase of a mere 2½ per cent. The additional wing loading is less than one pound per square foot, so manoeuvrability is barely affected. You can reduce the rate of climb by 2½ per cent if you wish, but it's hardly worth it.

For a real change though, how about a turretless Defiant, with 4 × .303-inch Brownings. Deduct 750 lb for the turret plus gunner, add back 20 lb for repairing the hole and 200 lb for frontal armament and you have a 530 lb total weight saving, bringing the Defiant into much the same category as the Hurricane. The drag saving by eliminating the turret has got to be near enough 25 mph on the top speed, and the new rate of climb could reasonably be assessed as half way between the original Defiant and the Hurricane. Combine this with a saving in wing loading of 2.21 lb per square foot and what have we now? Well, we could certainly have a go at a Bf 110 in it.

Chapter 5

Observation

To sit in the cockpit of a World War 2 piston-engined fighter is to be aware of how much you can't see. Straight ahead of you is a large engine cowling effectively blocking the view forwards and downwards. On each side is a wing large enough to conceal a whole squadron of aircraft half a mile away, while the fuselage and tail also obscure quite a hefty slice of landscape. As if this isn't bad enough, the canopy frame is thick enough to completely cover a fighter at a mere five hundred yards range.

One needs to see the enemy in order to shoot him down; in addition one needs to see him in order to avoid being shot down by him. In the 1939-45 era, the pilot's most essential piece of equipment was the Mark 1 human eyeball, the effective use of which must be considered first. Formations can be seen at long distance more easily than single aircraft. The larger the formation, the more easily it can be spotted. A moving object is more easily picked out than a static one so, for example, an aircraft taking off would be easier to spot than a parked one. Colour contrast makes aircraft easier to spot; you would see one silhouetted against a cloud far more easily than an aircraft against the ground. An aircraft flying close to the angle of the sun is very difficult to see, as you would be dazzled peering into the sun to look for it. Finally, there are the blind spots caused by the aircraft you are flying, which can only be checked by adopting a weaving mode of flight, constantly lifting your wings and tail out of the way while you peer anxiously beneath them like a nervous auntie at bedtime looking for burglars.

My preferred method of simulating observation is to use the clock code, but in

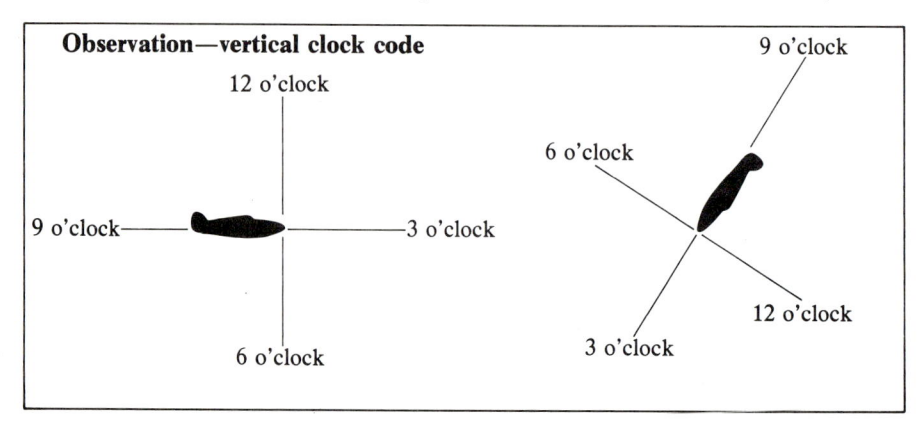

Observation—vertical clock code

Observation 41

the vertical rather than the horizontal plane. 12 o'clock is vertically above the cockpit. Three o'clock is directly in front of the nose. Six o'clock is vertically below the cockpit, and nine o'clock is directly astern. This applies regardless of the attitude of the aircraft, see sketch.

The best, or least obstructed area of observation is that between 12 and three o'clock, fairly closely followed by the sector 12 to nine o'clock. Next comes the sector three to six o'clock, with the sector six to nine o'clock distinctly the worst. We are, of course, assuming a complete search of the sky in every move. We can, therefore, assess the percentage probability of spotting a single aircraft in these sectors per move as follows: 12 o'clock to 3 o'clock—60 per cent; 3 o'clock to 6 o'clock—40 per cent; 6 o'clock to 9 o'clock—20 per cent; and 9 o'clock to 12 o'clock—30 per cent.

In the event of a dispute, the smaller percentage is taken. The percentage probability increases by one and a half times when the range decreases to 60 cm or less, ie, the probability of spotting an opponent in the 6 o'clock to 9 o'clock sector would increase from 20 per cent to 30 per cent at close range. Should an opponent be silhouetted against a cloud, this should make him an extra ten per cent easier to see, ie, 30 per cent would become 40 per cent. If, however, he is within ten degrees of the angle of the sun, the probability would be divided by four, ie, 60 per cent would become 15 per cent. Furthermore, if a pilot is engaged with a hostile aircraft, either attempting to close with it or attempting to get away from it, the probability of observing any other aircraft is also divided by four, as his attention will, to a great degree, be focused upon his immediate opponent. If he is in position to fire at anything, this will take all his concentration, and he will be completely oblivious to all other aircraft except the target unless he is also being shot at. Anyone being

The difficulties of a head-on attack are illustrated here. The aircraft is a B-17 Flying Fortress, one of the largest aircraft to participate in World War 2. The image size in this picture shows it as it would appear at 250 yards' range. An attack would generally be commenced at 1,000 yards, opening fire at 600 yards, when the apparent size would be barely half. At a closing speed of 500 mph, 600 yards would give less than 2½ seconds. At 250 yards' range, you would be fractionally more than one second from a collision (*Minyon Prescott*).

This picture gives an image size of 100 yards. The B-17 looks big enough to hit, but you have less than half a second left. At this range, a Bf 109 would be less than one third of the size (*Minyon Prescott*).

fired at automatically becomes aware of the firing aircraft when he sees tracer flashing past. Friendly aircraft can, of course, be warned by radio, but this takes one full move. They must also be in the same unit (squadron or wing, *staffel* or *gruppe*, etc) so that they can be assumed to be using the same radio frequency.

The clock code method of observation is fine for single-seaters, but complications arise with multi-crewed aircraft. The way to overcome this is to allot a sector of sky to each crew member able to keep a lookout, bearing in mind that in some heavy bombers crewmen such as the navigator and wireless operator are unable to do this, also that whilst on a bombing run, the bomb-aimer will be totally pre-occupied, and the pilot only 50 per cent effective as he will be busy following the bomb-aimer's instructions. In any case, evasive action while on a bombing run is impossible without spoiling the aim. Sectors of observation should be allocated with 3 o'clock directly ahead of the crew member according to whichever way he is normally facing. This means that in a two-seater such as the Bf 110c, responsibility for search is split into two sectors. The field of search for the pilot is 12 to 3 o'clock and 3 to 6 o'clock, the rear gunner covering the arcs 6 to 9 o'clock and 9 to 12 o'clock. As, however, the gunner is facing rearwards, his view is better than his pilot's and his probabilities become 40 per cent for the sector 6 to 9 o'clock and 60 per cent for the sector 9 to 12 o'clock.

In multi-crewed aircraft such as the Lancaster, the position is a little more complex. What we have to do is assess the visibility from each position, together with the percentage probability of sighting. The nose gunner's position has excellent visibility both up and down as well as straight ahead. Anything behind him is not really his responsibility, and he confines himself to his allotted sector of sky. The rear gunner does exactly the same but in reverse. The pilot has much the same sort of view as in a single-seater. It is arguably worse downwards, but as we are trying to legislate for all multi-crewed aircraft, let's keep it simple and call it the same. The mid-upper gunner has the grandstand seat, and would act as co-ordinator in the event of an attack. He can face either way with equal facility. Beam gunners such as were carried in the B-17 are awkward to account for satisfactorily, but the easy method is to say that two beam gunners equal one mid-upper gunner. Two sorts of ventral gunners remain to be accounted for. The first is in a ventral turret. The position is cramped and visibility is poor. The second is in a gondola, usually facing rearwards, with limited visibility and a restricted arc of fire, and with a good possibility of being trampled on by an excited beam gunner, as in the Heinkel 111H. A final point to remember, which also applies to the Heinkel, is that the nose gun is operated by the observer, whose function it also is to aim the bombs and navigate. Therefore, when on a bombing run, no nose gunner in some aircraft.

Observation with multi-crewed aircraft is, therefore, assessed as follows:

Percentage probabilities of observation

Position	Sectors (clock code)				Position	Sectors (clock code)			
	12 to 3	3 to 6	6 to 9	9 to 12		12 to 3	3 to 6	6 to 9	9 to 12
Nose gunner	60	60	Nil	Nil	Rear gunner	Nil	Nil	60	60
Pilot	60	40	20	30	Ventral turret	Nil	50	50	Nil
Mid-upper	60	30	30	60	Ventral gondola (facing rear)	Nil	Nil	50	Nil
2 beam gunners	60	30	30	60					
1 beam gunner	30	15	15	30	Ventral gondola (facing front)	Nil	50	Nil	Nil

It can be seen that on many occasions more than one crew member will be able to spot an attacker. Where this occurs, and a multiple attack takes place, the dice will be thrown for all possible sightings. The order for this will be the nearest attack first, followed by the rest in distance order. No crewman can make more than one sighting per move, but a hysterical yell down the intercom will instantly alert the others.

To finish with observation, it is sufficient to sight an enemy once unless it vanishes into cloud, when the whole performance of spotting has to be repeated. The assumption is that, once having sighted the enemy, the pilot will, even if concentrating on other things, retain an awareness of the whereabouts of his opponent.

Terrain is the final subject of this chapter, which may seem a bit odd in an air game. Land battles have woods, rivers, hills, etc, which are basically things to hide behind, then when the enemy goes past, you jump on him and say 'gotcha!'

In the air you have two things to hide in; the sun and clouds. The sun, as stated earlier, dazzles and is very easy to hide in and should only be omitted from the game in exceptional circumstances. An indication of the importance of the sun is that it was plotted on Fighter Command Operations Tables from quite early on in the war.

Unless you are fighting very near the equator, the sun will never be overhead. You can, if you wish, calculate an angle for the time of day, but I find it just as satisfactory to roll a normal (six spot) dice and use the following diagram.

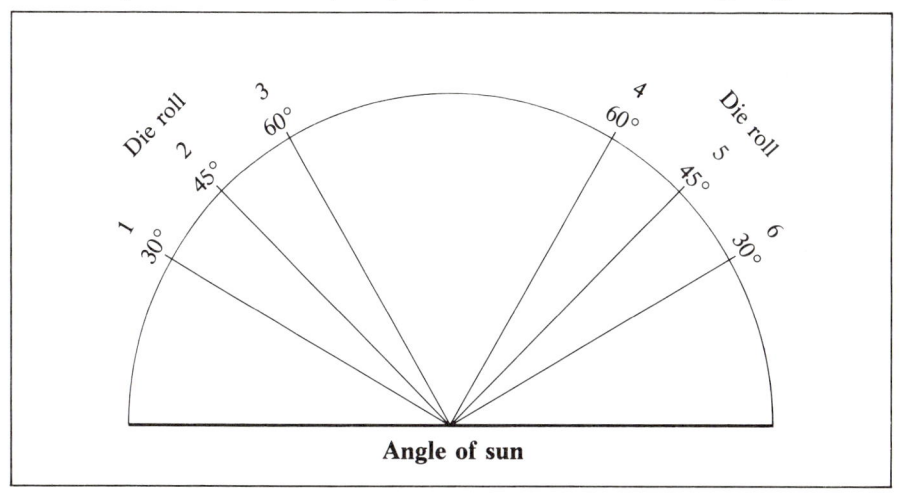

Angle of sun

Cloud is the other form of cover. Artistically cut from polystyrene ceiling tiles, they should be propped up above the playing surface at a sufficient height to allow the models to hide under them. Just two or three will be quite sufficient and can add a lot of interest to a game. A point here is that, when you are in cloud, both you and your opponent know roughly where you are. The best way of preserving the integrity of a cloud is to write orders which take you clean through and out the other side before you enter it. Failure to do this means that you must straighten out at the point where your written orders end and keep going until you finally do break cloud, continuing in a straight line until the end of that move.

Chapter 6

Aerial gunnery

It is necessary at this stage to examine the weaponry of the period. In 1939, the most common weapon was the rifle calibre machine-gun. Its advantages were a high rate of fire, and a relatively light weight so that more of them could be carried, thus giving a relatively large pattern of fire which tended to give a greater chance of a few hits to the mediocre marksman. The heaviest weapon in general use at this time was the 20 mm cannon, with a slower rate of fire. This gave less chance of a hit, but the hits were far more damaging. A compromise between the two was the heavy machine-gun, whose rate and weight of fire was between these two extremes. Other air-to-air weapons were used, such as the American 37 mm cannon fitted to the Airacobra and the German 30 mm MK 108, and BK 5 of 50 mm calibre. In addition, air-to-air rockets of various types were used and experiments carried out with air-to-air bombs.

As the war progressed, aircraft were progressively up-gunned, the favourite weapon being the 20 mm cannon except for American aircraft, which tended to stick with the 0.5-inch Browning heavy machine-gun. At this point it will be helpful to examine and compare the performance of the various main types of gun used. The weapon favoured by the Armée de l'Air, and later by the Royal Air Force, was the Hispano Suiza. The rate of fire was 0.0856 seconds per shell which, when evaluated against its muzzle velocity of 2,820 feet per second, means a gap of over 80 yards between shells at close range. An aircraft 30 feet long, crossing at a 90 degree angle at a speed of 300 mph, would be lined up for a mere 0.0681 seconds and, although from this it can be seen that the odds favour a hit, the percentage probability of scoring one single hit from one gun, even assuming that the aim was accurate, was only 79½ per cent. Finally, the Hispano in its earlier versions were fed by a 60-round drum, which gave only 5.14 seconds firing. Later, of course, belt feeding was introduced which gave much longer periods of firing.

Compare these figures to the .303-inch Browning as used in the early Spitfires and Hurricanes. The rate of fire was 0.0444 seconds per bullet, or nearly twice as fast as the Hispano. The muzzle velocity of 2,400 feet per second gives a gap of only 35 yards between bullets; an aircraft flying as in the previous example would be lined up for 0.0681 seconds, but given a true aim the chance of one gun hitting would be one certain hit and a 53.38 per cent probability of a second hit being scored. In other words, in deflection shooting of any sort, the rifle calibre machine-gun was, gun for gun, about twice as effective in scoring hits. Combine this with the fact that two .303-inch Brownings would cram into the space occupied by one Hispano, and it can be seen that the Browning is potentially four times as effective

Aerial gunnery 45

at scoring hits. Despite this, the Hispano displaced the Browning as the main fighter weapon in the Royal Air Force, the reason being, of course, that one hit from a 20 mm shell would do much more damage than four hits from .303-inch machine-gun bullets, which were relatively ineffective against armour and self-sealing fuel tanks. There are several instances of German bombers reaching home in 1940 with upwards of 200 hits from the small stuff. Divide this by four and the result would be 50 hits with 20 mm, after which any resemblance to an aeroplane would be pure coincidence.

There are two main ways of approximately evaluating the fire effect of different weapons. One is to calculate the sheer weight of metal hurled at the enemy in a give period. The other is to calculate the energy transmitted, ie, weight of metal × muzzle velocity converted into foot-pounds (or Newtons if you wish to be modern). The second way has a couple of built-in banana skins. The first is that muzzle velocity is exactly what it says; the velocity on leaving the muzzle of the gun. To illustrate this let us take the German MK 108 as an example. Its muzzle velocity was 1,750 feet per second, or 1,193 mph. However, the time taken for the shell to cover a distance of 1,000 yards was slightly in excess of two and a half seconds, an *average* speed of barely 800 mph and, in fact, at this point, the shell would be travelling at about half this speed, or 400 mph. Between the muzzle and a point 1,000 yards away, the speed variation would be in the region of two-thirds. As you can see, this would alter the energy transmission figure by far too much to be usable for our purpose.

The second banana skin is relative speed. A head-on attack could mean a closing speed in excess of 600 mph, while it would not be beyond the bounds of reality to postulate firing at a target which is diving away at a speed 200 mph faster than your own aircraft. This would give a possible speed variation of about 800 mph, which would give an energy transmission calculation varying by two-thirds about the norm. This means that for the needs of the wargame we are left with assessing the fire factor by the weight of metal. There is a ballistic reason for doing this anyway. A projectile striking home usually hits the skin of the aircraft first and passes through, leaving a small neat hole. However, unless it met the skin precisely at right angles it would be deflected from its straight path slightly and this would be enough to make it topple, or start going end over end. The next hole it makes is likely to be large and ragged. Therefore, while the difference between small neat holes made by 7.62 mm and 12.7 mm bullets may not seem very significant, the large ragged secondary holes would show an appreciable difference.

The other factor to consider before we compare weights of fire is ammunition. Machine-guns generally contained a mixed loading of ball, armour piercing and incendiary. Cannon loadings would usually contain a proportion of high explosive also, which on impact would cause a great big hole. So much so that a single hit from a 30 mm cannon would be enough to blast most single-seaters or even light bombers clean out of the sky.

Now for a comparison of weaponry. The rifle calibre machine-guns did not vary overmuch in performance, and an average weight of fire would be in the region of 0.40 lb per second. The heavy machine-guns likewise were all in approximately the same performance bracket and a fair average would be 0.90 lb per second. The big variation comes when we consider cannon, and for this I have prepared a table, which also includes a suggested fire factor using the rifle calibre machine-gun as a base with a factor of one.

Type of gun	User	Weight of fire per second	Fire factor
Rifle calibre machine-gun	All	approx 0.40 lb	1
Heavy machine-gun	All	approx 0.90 lb	2
20 mm Hispano-Suiza	Britain USA France	3.21 lb	8
20 mm Oerlikon MGFF	Germany Japan	1.28 lb	3
20 mm Mauser MG 151/20	Germany Japan	2.73 lb	7
20 mm ShVak	USSR	2.92 lb	7
30 mm Rheinmetall-Borsig Mk 108	Germany	7.91 lb	20

To assess the total fire factor of types of aircraft, proceed as follows:

```
Spitfire V       2 × 20 mm Hispano:         2 × 8 = 16
              + 4 × .303-inch Brownings:    4 × 1 =  4
                                                   20 points
FW 190A-3        2 × 20 mm MG 151/20:       2 × 7 = 14
              + 2 × 20 mm MGFF:             2 × 3 =  6
              + 2 × 7.9 mm MG 17:           2 × 1 =  2
                                                   22 points
```

You may feel at this point that it is wrong to lump together the combined fire effect of weapons of differing muzzle velocities and consequently variable effective ranges. Personally I feel that it is valid to combine the total fire factor as, while the weapons may vary, the pilot's aiming ability will be constant, and few could hit anything at more than 300 yards anyway. Cannon could, in most cases, be fired separately to machine-guns, but the general tendency with a target in your sights would be to let fly with the lot. The thing you must not do, of course, is to lump the bomber's defensive fire together. Each gun position aims and fires individually.

The other problem is that of ammunition supply, which can take some keeping track of. In the example previously, the FW 190A-3 had three different types of guns. The MG 17s were supplied with 1,000 rounds per gun, giving about 50 seconds firing time. The MG 151/20s had 200 rpg which, at 750 rounds per minute, would last for 16 seconds, while the MGFFs outboard in the wings were fed from 60-round drums, which lasted ten seconds. The early Hispanos were also fed from a 60-round drum, and were in much worse case with their higher rate of fire. A drum lasted barely five seconds. I would suggest here that, for an interesting game, allow everyone to have 12 seconds worth; this discourages optimists from blazing away on 1,000 to one chances and is quite adequate for normal use.

One final point; some bombers had guns fitted with saddle magazines. Allow three magazines per gun, four seconds' firing per magazine, with one complete move to change the empty magazine for a full one.

There were, of course, larger air-to-air guns mounted than we have dealt with so far. A classic example is the German BK 5 of 50 mm calibre which fired a shell weighing 3½ lb at a rate of one very 1⅓ seconds. One hit would pulverise anything that ever flew, so that there is no need to quantify a fire factor for this gun, or

Aerial gunnery 47

indeed any gun larger than 30 mm. One hit equals one kill, although with such a slow rate of fire the chance of a hit must be halved.

One final point before we leave guns and examine the art of gunnery, and that is recoil. When a large number of guns, particularly heavy ones, were fired it caused a recoil which had a distinct braking effect upon the aircraft. For example, the Henschel Hs 129B3, with its massive 75 mm anti-tank gun firing shells weighing more than 26 lb, was even reputed to have given the occasional pilot a hernia.

To include recoil effect is, of course, not strictly necessary to the game but it certainly livens things up when a player is firing from an aircraft close to stalling, and in this context will often mean the difference between a player firing a three-second burst or a one-second burst. An approximate calculation can be made as follows: total weight of fire per second multiplied by the muzzle velocity equals the force, or energy, exerted on the firing aircraft in the form of recoil. Divide this force by the normal laden weight of the aircraft and you will have the distance that the aircraft should in theory be moved by that amount of force, expressed in feet. Recalculate this in terms of miles per hour rounded off to the nearest whole number and this will be your speed loss caused by the recoil of one second's firing.

Before anyone writes to me with a complex mathematical formula in one hand and a big stick in the other, let me say that I know that things are not this simple, but it is an easy way of arriving at an answer which gives reasonable comparative differentials. And just to really show you what I think of it, to make it worth using in a game, enhance it by doubling it. Climb and dive have been exaggerated to speed the game up; why not recoil also? One final word of warning before I give a worked example of recoil; do not attempt to use it for swivelling guns, as it then becomes far too complicated.

Now for an example. Let us take a Hurricane II armed with four 20 mm Hispano Suiza cannon.

Weight of fire per second (3.21 lb) multiplied by muzzle velocity (2,820 feet per second) multiplied by the number of guns (4) = 36,209 foot pounds of energy. The normal loaded weight of this fighter was 7,470 lbs. Energy divided by weight = $\frac{36,209}{7,470}$ = 4.85. This means that the energy transmitted in recoil in one second would push the Hurricane back 4.85 feet. 4.85 feet per second is 3.31 mph. Double this for enhanced effect and round off = (6.62) 7 mph. A three-second burst would then slow the Hurricane by 21 mph which is well worth including in the game.

Having examined the guns in some detail we now need to assess their effect. Some information is available but very little and fairly non-specific. What we must do is apply common sense in the interpretation of what data is available, also the word 'average'. We know, for example, that it is only common sense that even the toughest aircraft can be shot down by a single rifle calibre machine-gun. Equally, aircraft have been known to stagger home after the most fearful hammering. Both these points must be borne in mind.

I do not propose to quote obscure instances or masses of data which would only confuse the reader, but will draw conclusions from information contained in two sources only, *World War II Fighter Conflict*, and *The Bomber in World War II*, both by Alfred Price.

Luftwaffe investigators concluded that, on average, three or four hits from a 20 mm cannon would destroy a single-engined fighter, 20 hits a heavy bomber.

Although it is not specifically stated, I am assuming that the Mauser MG 151/20 would generally be the cannon referred to. The MK 108 cannon of 30 mm was much more effective and, I quote: 'Its 11 ounce incendiary or high explosive shell had an effect on aircraft structures which was truly devastating, against fighters *or medium bombers** a single hit was almost invariably sufficient to cause their destruction; against a four-engined heavy bomber, three or four hits were usually enough.'

We are also given some tabular information which I quote in full, together with the lead-in.

'One of the most detailed sources of information on the relative causes of aircraft losses is from a US Navy survey of aircraft lost or damaged by Japanese air-to-air action between September 1944 and August 1945. The 501 single-engined aircraft considered in the survey include torpedo bombers and dive bombers as well as fighters; but since all were of similar layout and construction, the figures are relevant to this account.

Single-engined aircraft

Position of hit	Total number of aircraft hit	Number of aircraft lost	Percentage loss of aircraft hit
Propeller	9	0	0
Power plant	37	23	62
Structure	215	23	11
Pilot and/or controls	97	74	76
Control surfaces	27	0	0
Oil system	27	23	85
Fuel system	30	24	80
Hydraulic system	35	21	60
Electrical system	6	0	0
Others	18	5	28
	501	**193**	**38**

'These statistics require some qualification before they can be considered representative for all fighters. Throughout the survey, a lost aircraft was defined as one that had failed to return to a friendly base after being hit; a damaged aircraft was one that had returned to base after being hit, whether it was repairable or not. All the US Navy aircraft in the sample were powered by air-cooled engines; had these aircraft been fitted with liquid-cooled engines almost all of those hit in the engine (or its associated cooling system) would have been lost. It is likely that the number of pilots hit was greater than the figures would suggest, since those aircraft that went missing without trace were not included'.

A similar survey was conducted with PB4Y Liberator bombers during the same period with very similar comments. The Liberator, having four engines, was not as vulnerable to engine damage unless fire broke out, being able to return on three or occasionally even on two engines. The structure was much larger with heavier members than a single-engined aircraft, and consequently less vulnerable. The pilot, if hit, could be relieved, consequently loss from this cause was potentially less. The oil system is directly related to the four engines with a resultant drop in vulnerability. The following table refers to 354 instances of loss or damage.

*My italics.

Aerial gunnery

PB4Y Liberator

Position of hit	Total number of aircraft hit	Number of aircraft lost	Percentage loss of aircraft hit
Propeller	7	0	0
Power plant	57	21	37
Structure	135	5	4
Pilot and/or controls	29	6	20
Control surfaces	20	0	0
Oil system	9	3	33
Fuel system	31	8	26
Hydraulic system	17	2	12
Electrical system	9	1	11
Others	40	0	0
	354	46	13

What does a comparison of these figures tell us? First, there is no definition of structure; if we re-classify the minor areas of damage under this heading as well as the intriguing heading 'others', and define damage to the structure as being holes in the aircraft in any other place than in the main damage areas, we arrive at the following:

Damage areas	Single-engined aircraft % Aircraft hit	% Loss of aircraft hit	Liberators % Aircraft hit	% Loss of aircraft hit
Engines	8	62	15	37
Pilot	19	76	8	20
Oil	5	85	3	33
Fuel	6	80	9	26
Hydraulics	7	60	5	12
Structure	55	10	60	3

From this we can prepare a table showing the probability of an aircraft sustaining damage in any given place, with straight interpolation for twin- and three-engined aircraft. The two items for which we have no data are injury to crew members, also the incidence of structural damage between wings, fuselage and tail. However, we may assume that the wings and tail have an equal chance of being hit, while the fuselage, being both bulkier and usually the aiming point, has twice the chance of the other two areas of sustaining damage. Hits on crew members are a trifle more difficult to judge. The percentage of structure hits varies little between single- and four-engined aircraft, but there is a wide divergence in the percentage of pilot/controls hits, which in any case seems to he inordinately high. For the sake of the game I would suggest that the percentage of pilot hits is halved thus ignoring hits on the controls and the balance of hits is added to the structure. To assess the remaining crew members accurately is not possible, but if we assume that a fuselage hit must be scored to injure a crew member, and that the injury must be sufficiently severe to seriously impair his fighting capacity, then I would suggest that the number of crew exclusive of the pilot be pro-ratad to the points value of the fuselage. For example, a four-man crew in a fuselage 20 points strong would be knocked out one by one at five point intervals. The particular crew member at risk would be decided by dice. This would give a ten per cent chance of the pilot being hit in a small aircraft and a four per cent chance in a large one.

Anyway, we can now prepare a hit probability table.

Hit probability table

	\multicolumn{8}{c}{Number of engines in aircraft}
	One		Two		Three		Four	
	From	To	From	To	From	To	From	To
Engines	00	07	00	09	00	11	00	14
Pilot/controls	08	17	10	18	12	20	15	22
Oil	18	22	19	23	21	24	23	25
Fuel	23	28	24	30	25	32	26	34
Hydraulics	29	35	31	36	33	37	35	39
Fuselage	36	67	37	68	38	68	40	69
Wings	68	83	69	83	69	83	70	84
Tail	84	99	84	99	84	99	85	99

This table is used when a hit has been scored. Roll the percentage dice. If the damaged aircraft is twin-engined, and the dice score is 35, this locates the hit in the hydraulics, the area 31 to 36 in the twin-engined aircraft column covering hydraulic damage.

Having assessed firepower, the location of hits scored and the percentage losses of aircraft hit in any given place, we now have to put the information together to assess fire effect. Before doing this, it is necessary to have a brief look at the aircraft being hit. The single-engined aircraft operated by the US Navy during the statistical period under discussion were protected just about as well as was feasible. The F6F Hellcat, for example, carried over 4 cwt of protection in the form of steel plate, toughened glass, dural deflector plates and self-sealing material for the fuel tanks. In addition, all aircraft in the sample were radial-engined. Had they been liquid-cooled, with their vulnerable plumbing, the percentage loss of aircraft hit in the engine would be something like half as much again, or 93 per cent. The Liberator is a rather different bird. The remarks on liquid-cooled engines obviously apply (the Liberator has radials) but, unlike the single-engined aircraft, an engine hit, unless accompanied by an uncontrollable fire, is unlikely to be lethal. At least two engines must be put out of action to be decisive. Looking at a cutaway drawing of this aircraft one is aware of the large voids inside it. There are so many places where a projectile could go straight through with no more damamge than a punctured skin that it is easy to see why the smaller aircraft are so much more vulnerable. The Liberator also has its fair share of protection for the crew and the fuel tanks. The figures that we have obtained appear to relate to exceptionally strong and well protected aircraft and would probably only be bettered by a few Russian or German ground attack aircraft. The Liberator seems remarkably well able to sustain damage compared to the single-engined aircraft. We must now try to assess what was hitting them, which, unfortunately, has not been recorded.

To do this it is necessary to make an arbitrary assumption which is that all hits were scored by four types of Japanese aircraft; as this is the Pacific theatre from late 1944 onwards, I propose to use the A6M5 Model 52 Zero-Sen, the Ki 84-Ia Hayate, the N1K2-J Shiden-Kai, and the J2M3 Raiden, the purpose of this being to assess a reasonable average of the firepower brought to bear which achieved the results in the USN survey. All these aircraft were cannon-armed, the Japanese Type 99 cannon, their most widely used air-to-air weapon, being a licence-built Oerlikon, and their Type 5 cannon a licence-built Mauser, both with performance comparable to the German originals. The fire factors, based on our armament calculations would be 10, 10, 12 and 12 respectively, giving an average fire effect of

Aerial gunnery 51

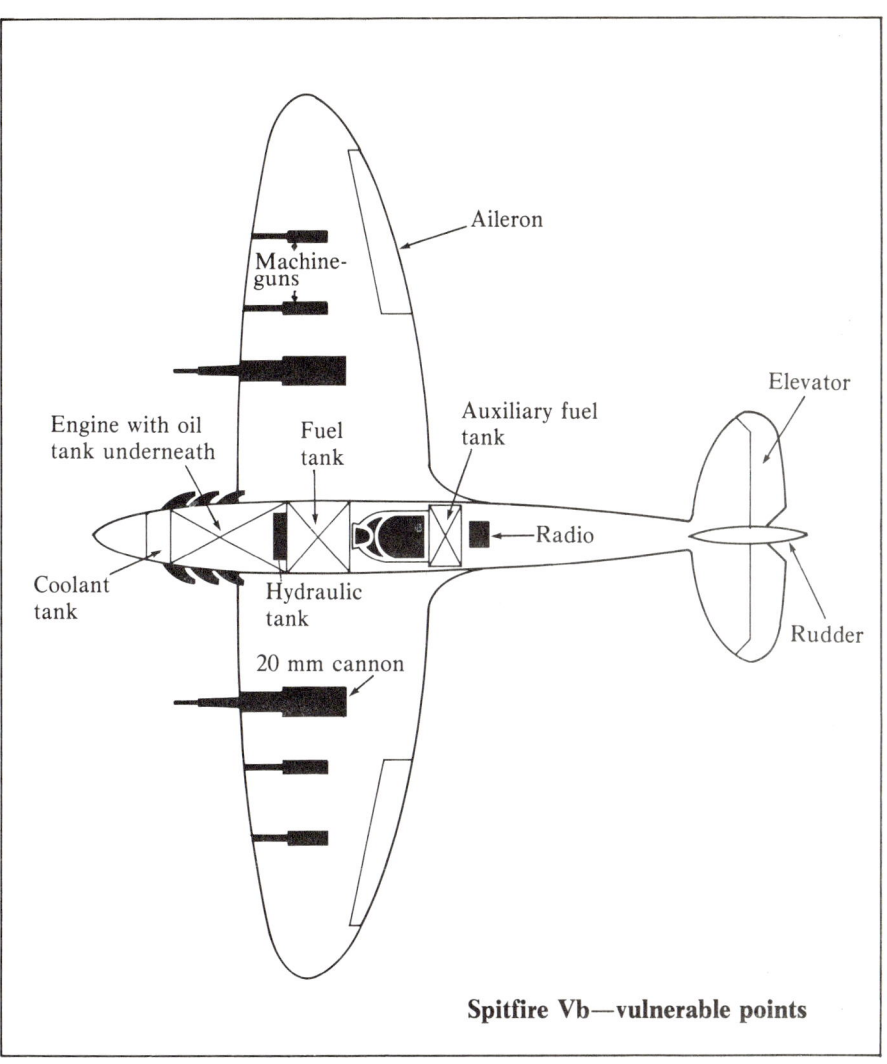

Spitfire Vb—vulnerable points

(Labels: Aileron, Machine-guns, Elevator, Engine with oil tank underneath, Fuel tank, Auxiliary fuel tank, Radio, Coolant tank, Hydraulic tank, Rudder, 20 mm cannon)

11 points per second. The other factor worth considering is that, in any one second burst in which hits are scored, rarely will *all* the projectiles strike home. The probability of how much does hit is capable of theoretical analysis subject to deflection, relative speed, and a large allowance for pure luck, but to use this would mean a large and complex chart and consequent slowing of play. We must, however, allow something for this and the simple way is to roll one percentage dice, multiply the number thrown by the fire factor and divide by ten, always rounding fractions upwards. For example, a dice throw of six on a fire factor of eight, divided by ten, would give a score of 4.8, rounded up to five.

To return to our comparison of data, the average fire factor is 11, and the average score seven, this being arrived at by combining the fire factor with the average die throw, and it is with an average score that we should assess the ability of aircraft to sustain damage and stay in the air. We will then first deal with engine hits.

Hits in the engine will generally have one of the following results: a) No apparent effect; b) partial loss of power; c) complete loss of power; d) fire; and e) disintegration (in extreme cases). Our data for single-engined aircraft shows that, of all aircraft hit in the engine, 62 per cent were lost. Lost in this case is defined as failure to return to a friendly airfield, not necessarily shot down on the spot.

Whilst I cannot substantiate the assumption, it seems reasonable to suppose that two thirds of the 62 per cent were shot down on the spot, the others falling by the wayside while attempting to reach home with engine damage. We must remember that the statistics are from the Pacific Theatre, where airfields were few, carriers often difficult to find, having shifted possibly 50 miles or more since take-off, and a great deal of flying was done over water which had few, if any, landmarks. Had the battles taken place over a friendly land area with airfields at frequent intervals the overall percentage of losses may have been rather lower. In addition, of those aircraft shot down during the battle by hits on the engine, a proportion would lose power and attempt to fight their way out before succumbing. If we arbitrarily assess this at one third of the aircraft shot down during the battle, we have a workable basis for assessment as follows, for single-engined aircraft sustaining engine hits.

1 Percentage of aircraft shot down on the spot—30.
2 Percentage of aircraft sustaining damage sufficient to stop the engine within, say, one minute—10.
3 Percentage of aircraft sustaining damage sufficient to stop the engine within, say, 30 minutes—22.
4 Percentage of aircraft sustaining minor damage and able to get home—38.

Returning to our average fire factor of 11 points it is obvious that if 30 per cent of aircraft are knocked down on the spot, it would theoretically (discounting lucky hits) require 70 per cent of the fire factor, or eight points, to strike home. We can, therefore, assess the ability of the radial engine to absorb damage at eight points, which is known as the defence value. When the defence value reaches nil, the engine is out of action. I have prepared a table based on the foregoing information giving the effect of various types of damage.

Radial engine—defence value eight
Defence value reduced to:
    4—minor damage—no apparent effect.
    3—moderate damage, engine loses half its acceleration factor, pilot goes home by shortest route.
    2—severe damage, engine loses all of its acceleration factor, pilot goes home by shortest route.
    0—engine stops, ie, same effect as throttle shut. Fire breaks out, throw one die per move, score 0 to extinguish. One extra damage point accumulates in each succeeding five moves, unless fire is extinguished.
    – 2—Uncontrollable fire causing two extra damage points per move. No chance to extinguish.
    – 6—Engine explodes.

As I have stated earlier, in-line engines were much more vulnerable than radials. If we arbitrarily assess the defence value of an in-line engine at seven points, it works out about right.

The results given above apply to single-engined aircraft, and must be adjusted for multi-engined aircraft as follows:
    Moderate damage to one engine—twin-engined aircraft lose one quarter of

Aerial gunnery

their acceleration, four-engined aircraft lose one eighth of their acceleration. Moderate damage to more than one engine is calculated in proportion.

Severe damage to one engine—twin-engined aircraft lose half their acceleration, four-engined aircraft lose one quarter of their acceleration.

Engine stops—twin-engined aircraft lose all acceleration, four-engined aircraft lose half their acceleration.

The next area to cover is the pilot. When hit, his natural tendency is to buzz off home sharpish. If badly hit, he will be unable to control the aircraft properly as his efficiency will be impaired. This, in the game, will have the effect of making his ability to turn the aircraft less, the aircraft will, therefore, turn one radius wider at all speeds. His shooting ability will be halved (the same effect will apply to all wounded air gunners) and his ability to spot attackers will be halved, (applies to all wounded crew members also). One final category remains—dead! In this case, the aircraft falls away vertically out of control and crashes. If a second pilot is carried he can take over after one clear move and recover an out of control aircraft.

The percentage losses of all single-engined aircraft with the pilot hit is 76 and we must remember once again the previous remarks about the Pacific Theatre. If we assume that, of the total losses, 50 per cent are knocked down on the spot, and 26 per cent (say 25 per cent) suffer impairment, we can say that to kill the pilot, 50 per cent of the fire factor (5.5 points rounded up to six) will be required. Our cockpit defence value can, therefore, be assessed as six points, although this needs arbitrary adjustment for unarmoured aircraft, and exceptionally heavily armoured aircraft, such as the Il 2, or the FW 190A-8, which carried something like 800 lb of protection.

Standard protected pilot—defence value six.
Heavily protected pilot—defence value eight.
Unprotected pilot—defence value four.

Defence value reduced to:
Any reduction—minor injury—pilot breaks off and goes home.
2—Serious injury—control, shooting and observation all impaired.
0—Killed—aircraft nose drops to 45 degree dive, crashes unless second pilot is available to recover.

Next in order is the oil system. Hits in this area tend to be more deadly than any other. The two main effects of hits in the oil system are fire, and the engine seizing up. As we need for realism a situation where an aircraft can be shot down with one point damage (a one second burst with a rifle calibre machine-gun) this is as good a place as any to work it in. If we say, albeit for the sake of the game, that a hit in the oil system invariably starts a fire which equally invariably engulfs a single-engined aircraft in two moves, but which a twin-engined aircraft has a 50 per cent chance to extinguish, a four-engined aircraft a 75 per cent chance to extinguish, although the last two both suffer the loss of an engine in consequence, we then have covered this David and Goliath situation.

We next come to fuel. This is very deadly in smaller aircraft but surprisingly less so in larger ones. The main danger is, of course, fire. It appears from our data that we need different defence values according to the size of aircraft, although we can reasonably assume that almost all aircraft lost due to hits in the fuel tanks or pipes went down at once as flamers. As the loss rate on single-engined aircraft is 80 per cent, we can assume that a fatal hit on a protected fuel tank in this type of aircraft can be caused by 20 per cent of the average fire factor, or 2.2 points, rounded up to three points. For the Liberator, 74 per cent of the fire factor will be required, which is 8.14 points, rounded up to nine points. We thus need a table for fuel hits, interpolating for twin-engined aircraft.

Self-sealing fuel tanks—defence values
  Four-engined aircraft—9 points.
  Twin-engined aircraft—5 points.
  Single-engined aircraft—3 points.
  If fuel tanks are not self-sealing, deduct one point from the above.

Defence value reduced to:
  0—Fire, causing one extra damage point per move.
  −4—Aircraft explodes.

Now to consider hits on the hydraulic system, which appear to be responsible for an astonishingly high proportion of losses on single-engined aircraft. The effects of hydraulic hits would tend to lie in three main areas: fire, loss of control, and unwanted bits flopping down and reducing performance. Hydraulic fire would not generally be very dangerous unless it spread to the fuel areas, and as a result I tend to discount fire as a main cause of loss. This leads me to consider loss of control to be the main bogey. Again though we have a wide discrepancy between the data for the single- and the four-engined aircraft, and unlike the engines and pilot, which have built-in adjustment factors, we need defence values according to size of aircraft.

Hydraulics—defence values
  Four-engined aircraft—9 points.
  Twin-engined aircraft—7 points.
  Single-engined aircraft—5 points.

Defence values reduced to:
  3—Undercarriage or flaps drop—acceleration and climb rate is halved, aircraft turns one radius wider at all speeds.
  0—Aircraft spins out of control—no recovery possible.

Finally we come to structure hits covering the areas wings, fuselage and tail. Here, our loss data shows that a large aircraft was only one-third as vulnerable to structure hits as a small one. If we then calculate on the basis as before, using the ten per cent structure loss for single-engined aircraft adjusted as before and divided arbitrarily into wings, fuselage and tail, we arrive at:

Wings and fuselage—defence values
  Four-engined aircraft—30 points.
  Twin-engined aircraft—20 points.
  Single-engined aircraft—10 points.

Tail—defence values
  Four-engined aircraft—15 points.
  Twin-engined aircraft—10 points.
  Single-engined aircraft—8 points.

NB: Japanese aircraft, all biplanes and militarily developed airliners, deduct one-fifth from defence values for wings, fuselage and tail.

Defence values reduced to:
  3—Aircraft stands a 50 per cent chance of breaking up if it reaches terminal velocity *or* turns on its tightest radius at any speed.
  0—Aircraft spins out of control—recovery impossible.
  −4—Aircraft breaks up.

We have dealt with armament and the effects of battle damage. We must now examine the problems of air-to-air aiming. There are plenty of books on the market

Aerial gunnery 55

dealing with the wartime experiences of the top scoring fighter pilots of all nations with the exception of the Italians, who did not encourage the 'ace' system and the keeping of individual scores. In many cases the 'dead-eye' marksmanship is almost monotonous, but we must remember that we are reading about the tiny minority of first-class shots, also that the story of their experiences has of necessity to be the highlights and does not usually include the many occasions on which they have spent several minutes milling about in a sky full of potential targets without once being able to bring their sights to bear. Three books in this category are, in my opinion outstanding. *The Big Show*, by Pierre Clostermann, seems to me to have more sense of immediacy than most; *Ginger Lacey—Fighter Pilot*, by R.T. Bickers, conveys graphically many of the problems of air-to-air aiming and, indeed, air fighting, which are enhanced by the sardonic humour of the subject; and *The Fighter Pilots*, by Edward H. Sims, covers the fighter pilot's war in more depth than any other and is more help in reconciling in one's own mind the enormous scores of the German fighter pilots when compared to the pilots of other air forces, than any other book I have yet come across.

A few words on this subject might be in order. Overclaiming was a feature of all air forces but it appears that in the overwhelming majority of cases this was done in good faith. There were many ways in which it could happen, stating them would only cause controversy and so I prefer to chicken out. Basically though, the reason for the comparatively high German scores was simply that they had more targets and flew more sorties. If you shot straight, and if you survived, you almost *had* to run up a large score. The same principle applied to the Japanese, but with decidedly inferior aircraft, survival was that much more difficult, although Hiroyashi Nishizawa was credited with 104 victories. However, for every rule there is an exception. The obvious one appears to be Hans-Joachim Marseille, credited with 151 victories in Africa. Whilst he obviously encountered a fair number of targets, his flying and shooting were so exceptional as to permit him to run up multiple victories while the remainder of his *staffel* were still wondering where to start.

The top scorer of any service was, of course, Erich Hartmann with 352 victories amassed in 1,425 sorties. This works out a approximately four sorties per victory, a far cry from the stories in the *Beano* of my childhood, but needs to be offset by the hundreds of German fighter pilots killed on the Eastern Front without scoring. Looked at another way, if Hartmann's multiple victory sorties could be offset by the Russian aircraft he damaged, this would mean that the top scoring fighter ace of all time flew no less than 1,073 fruitless missions.

By comparison, the allied top scorer in Europe, Johnny Johnson, flew 515 sorties for 38 victories, but a glance through his book *Wing Leader* will show how relatively few his opportunities were. On the other hand, if you really want a 'Dead-eye Dick', look no further than Oberleutnant Kurt Welter who, at the time that the USAAF was staging what almost amounted to a fly-past over Germany, scored 14 victories in his first 11 sorties, although I ought really to mention that he was shot down nine times during this period. After this he slowed up a bit, and only shot down a further 19 in his next 29 sorties.

These, then, were the sort of results which, with a lot of luck, could be attained. My purpose in stating this is simply to say that an average of one victory per game is far from being theoretically impossible, assuming that sufficient targets turn up, and in a game they will. All you have to do is to survive, and aim straight.

To return to aiming, the gunsight in general use during World War 2 was the reflector sight, which did not vary much between nations. It consisted of a reflector glass mounted in front of the pilot on to which was projected a circle of light with a spot in the centre, which the pilot did not see unless his head was in the correct position. In the event of electrical failure, the old ring and bead system was retained as a backup by some air forces. Certain individual pilots made a scratch on the windscreen in others. Whether the scratch really worked I don't know. Late in 1943, the gyroscopic gunsight made its debut with the British and American air forces. Combat analysis seems to indicate that it doubled the accuracy of aim. This may seem a lot, but if you consider overall fire effect would double from two to four per cent, it can be seen that the effect, although worthwhile, was not startling. The average pilot could just about hit a hangar if he was inside it; the gyro gunsight did assist him in deflection shooting, which otherwise would have been an utter waste of time. Denis Barnham, in *One Man's Window* described lining up carefully on an FW 190 at full deflection, firing, and hitting the '190 leader's wingman who was presumably somewhere astern, while on television recently Lieutenant Thomas G. Lanphier, who shot down Admiral Yamamoto's transport aircraft with his P-38 Lightning, stated that he fired a short burst to clear his guns before making an attack. To his amazement he hit the Ki 21 and set it on fire. This again was a full deflection shot which Lanphier, an experienced pilot, had no intention of attempting.

The factors affecting accurate aiming, apart from natural ability, are range, deflection, and relative speed.

First we will consider range. The further away from the aircraft you are, the smaller it appears. On one of the few wet days of 1976 my photographer and I visited the Imperial War Museum at Duxford, and among other things we examined the B-17G Fortress (privately owned by Euroworld) from a series of measured angles and ranges. At 600 yards, the maximum practical firing range with cannon, it had shrunk to insignificance, so much so that I distinctly remember saying 'where the hell is it!' Yet a head-on attack at 300 mph with the bomber moving at 200 mph would take slightly less than 2½ seconds from this distance.

What we have to do is to arbitrarily fix critical ranges and equally arbitrarily assess the chance of a hit at these ranges. You can if you wish calculate comparative sizes of targets at different ranges, working on the assumption that anyone could hit a Sunderland from 50 yards. The trouble is, if you do this with 100 per cent probability of hitting the Sunderland from 50 yards, the probability of hitting a Bf 109 at the same range is calculable at a mere 34 per cent, and this with no deflection or speed differential. While excellent for realism, it makes a dull game shooting down nothing at all in a fighter-versus-fighter encounter. Having done a considerable amount of trial and error, the word trial covering both games, patience and friendships, I have arrived at the following ranges and probabilities. If the choice of ranges seems odd, remember that our ground scale is one inch (or centimetre) = 40 yards, and that range in this game is defined as from the nose of the attacking aircraft to the *nearest point of the fuselage* of the target. I stress this to prevent argument, and wargamers are an argumentative lot generally.

What about the P-38? Certainly sir, it appears to have two fuselages. Prevent arguments by taking the nearest. I have in the past measured ranges from nose to nose, but at very close range this means that the models are playing piggyback which is unaesthetic to say the least. Any overlap at all means that the attacker has overshot and has not got a target. The necessity for this is caused by the fact that, compared with the distance scale, the models are vastly oversize, which has also

Aerial gunnery 57

made me add nearly another two-thirds to the ranges as part compensation. Played to true scale ranges, the game just does not look right. This is yet another example of exaggeration for the sake of the game. But back to business.

I have based ranges on a maximum effective of 1,000 yards, which becomes 25 centimetres for the tinies and 25 inches for the big boys, and, dividing this into ten equal parts, have calculated a parabolic curve starting with 95 per cent probability at minimum range, and one chance in 100 at extreme range. This gives much better accuracy than was the real-life norm, but we gamers are all crack shots, aren't we?

The next consideration is deflection. This is the angle subtended by the course of the attacker and the course of the target, as in the accompanying diagram showing a 30 degree deflection shot.

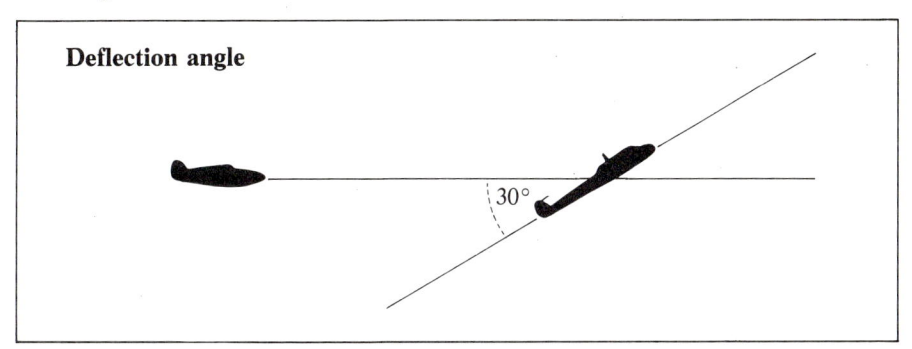

Deflection angle

In practice you would be aiming ahead of the target aircraft as your bullets would take a little time to cover the intervening distance and in this short time the target would have moved on. For example, let us assume that you try a 90 degree deflection shot from 300 yards and your bullets take two-thirds of a second to cover this distance. In this time, a target moving at 400 mph would have travelled 131 yards, and this is the amount of aim-off you would need to get a hit at this speed and range. As aim-off is infinitely variable we must ignore it and simply state that your sights are on when the centre-line of your aircraft can be extended to pass through any part of the target. This applies to fixed armament; swivelling guns are, of course, much more flexible. Deflection does, therefore, have a strong influence on accuracy, and I assess it as follows:

Deflection angles of up to 15 degrees are counted as no-deflection shots, and are on the same basis as range, being totally unpenalised. Angles of 16-45 degrees are penalised by two-thirds, for example 95 per cent becomes 32 per cent, and angles of 46-90 degrees by two-thirds again, ie, 32 per cent becomes 11 per cent. One final point before we leave the subject of deflection is that the description so far given is

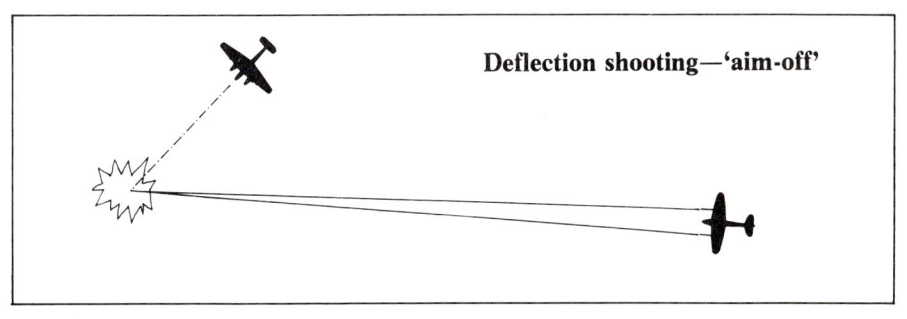

Deflection shooting—'aim-off'

applicable to a pilot firing fixed forward armament. A gunner serving a swivelling gun is, of course, in different case because he can often face directly into an attacker for a no-deflection shot. Deflection in this case is measured by taking the difference between the gunner's line of fire and the centre-line of the attacking aircraft.

This leaves comparative speed to assess. Obviously, the most accurate gunnery will take place from close range, no deflection, and the same speed as the target. The problem is getting to close range initially. If we start an attack from, say, 600 yards range and close to 100 yards at a speed 50 mph more than the target, we will take over 20 seconds, which gives plenty of time to line up the sights, check our tails in the mirror, check the turn and bank indicator to make sure we are not side-slipping, and still fire several bursts. At 150 mph we have less than seven seconds which is not nearly so good, and at 300 mph just over three seconds to do all these things, which is pushing it a bit. Speed differential then can be assessed at up to 50 mph, no effect; 51-150 mph, half effect; 151-300 mph, quarter effect; and speeds exceeding 300 mph, one eighth effect. We can now produce our gunnery table which is the percentage probability of scoring a hit in any one second of firing.

Gunnery table

Speed Differential	Deflection	\multicolumn{9}{c}{Range in millimetres}
		25	50	75	100	125	150	175	200	225	250
Not above 50 mph	0-15°	95	93	91	88	83	75	64	47	21	00
	16-45°	32	31	30	29	28	25	21	16	07	00
	46-90°	11	10	10	10	09	08	07	05	02	00
51 to 150	0-15°	47	46	45	44	41	37	32	23	10	00
	16-45°	16	15	15	15	14	12	11	08	03	00
	46-90°	05	05	05	05	05	04	04	03	01	00
151 to 300	0-15°	24	23	23	22	21	19	16	12	05	00
	16-45°	08	08	08	07	07	06	05	04	02	00
	46-90°	03	03	03	02	02	02	02	01	01	00
Above 300 mph	0-15°	12	11	11	11	10	09	08	06	02	00
	16-45°	04	04	04	04	03	03	03	02	01	00
	46-90°	01	01	01	01	01	01	01	01	00	00

This is the basic gunnery table, the following amendments are necessary. Rifle calibre machine-guns cease to be effective at 150 mm, heavy machine-guns at 200 mm. Firing cannot be effective beyond these ranges with these weapons. This restriction, which I freely admit is artificial, is to keep the gunnery simple. An aircraft with multiple type guns, say the P-40 with both light and heavy machine-guns, can use the combined fire factor of all its guns up to and including 150 mm range; above this the fire factor will be that of the heavy machine-guns only up to their maximum range of 200 mm. Alternatively, a Bf 109E, with light machine-guns and cannon, can fire at full effect up to 150 mm, but has the fire factor of the cannon only at greater ranges.

Swivelling guns were not as effective as fixed armament. If mounted in power-operated turrets they were not too bad, but if manually operated, the gunner had to struggle with the slipstream which reduced his efficiency no end, added to which his firing position was often not of the best. Alfred Price, in *The Bomber in World War II*, gives excerpts from reports on operational trials carried out by the Air Fighting Development Unit of the Royal Air Force. The following extract concerns the Heinkel 111H.

Aerial gunnery 59

'The beam gunner has to kneel on the window ledge or stand astride the fuselage, as his only alternative is to stand on the lower gunner, which no doubt he frequently does in moments of excitement'.

Let us, therefore, say that power-operated turrets and remote control barbettes have three quarters the table probability of scoring a hit, while the probability of hitting for manually operated swivelling guns should be halved. All fractions to be rounded upwards.

The maximum length of burst fired during a move would generally be three seconds unless it could be a continuation burst from the previous move, when four seconds can be taken.

At deflection angles exceeding 45 degrees, the maximum hits which can be scored in one burst is two, and at relative speeds exceeding 300 mph the same applies. Where the relative speed exceeds 300 mph and the deflection exceeds 45 degrees, only one hit per burst can be scored. In the event of the relative speed exceeding 700 mph, no-one has time to shoot.

The final item to cover before we leave this chapter is the calculation of relative speeds. This, in fact, is amazingly simple. Get hold of the largest piece of graph paper that you can obtain and carefully draw out the diagram illustrated here.

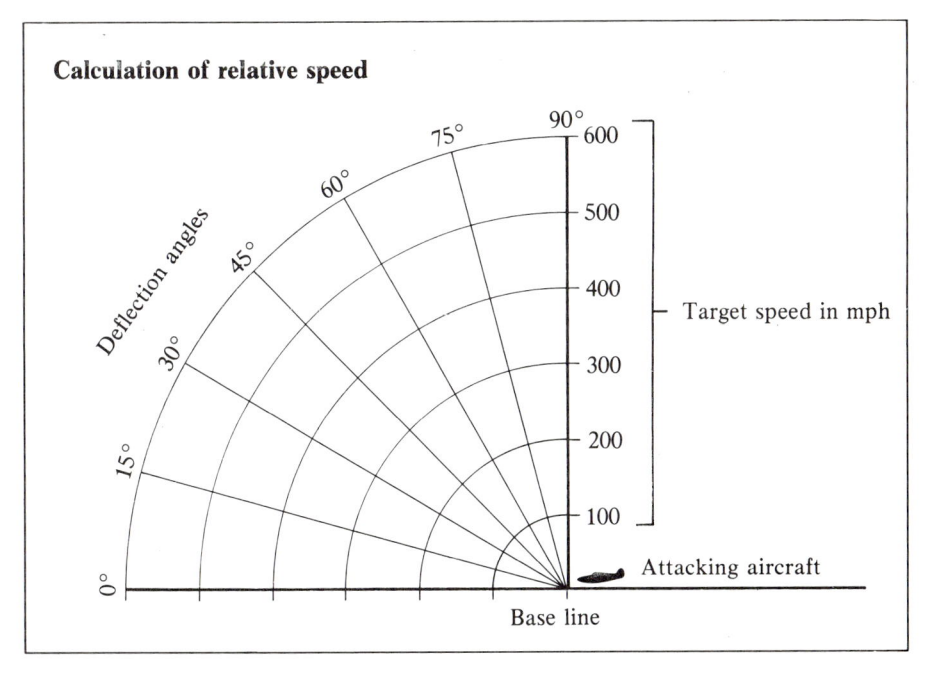

Calculation of relative speed

This can be used to give precise relative speeds. The speed of the attacking aircraft, say 327 mph, is recorded first. Let us assume that the target aircraft is going away at 420 mph at a 60 degree deflection angle. The rings on the diagram represent target speeds in mph. Follow the 60 degree line out to 400 mph, then measure out another 20 mph. From this point, drop a vertical line to the base line and read off the relative speed of the target, in this case 213 mph. This is slower than the speed of the attacker so deduct the speed of the target from the speed of the attacker, which gives a closing speed of 114 mph. If the target had been coming towards the attacker, the relative speed would be added to give a closing speed of

540 mph. Had the relative target speed been greater than that of the attacker, it would have given an opening speed, which is a closing speed in reverse. In the event of a zero deflection angle, the relative speed is the difference between the speeds of the two aircraft if both are going in the same direction, and the sum of the speeds if they are approaching head-on. In the case of a 90 degree deflection target, the relative speed is that of the attacking aircraft only.

We have now covered the most usual way of shooting down the enemy, and a brief look at the more unusual ways is in order.

Air-to-air bombing was used by the Germans and Japanese against the large American bomber formations. The attempts by the Luftwaffe were initially on a rather *ad hoc* basis and involved such experiments as dangling bombs under the aircraft on lengths of wire. As the bomb tended to stream out behind rather than hang vertically, this was not a success. This was followed by attempts to drop bombs directly on top of the enemy formation. One had to fly a specific height above the bombers (the bombs were time fuzed, otherwise the Fatherland would have been plastered by its sons), synchronising one's speed to that of the bomber formation, which was fairly easy if no escort fighters were around except for the little matter of judging the height separation correctly. A 500 lb bomb had to explode within 100 yards laterally or above a bomber to be effective; if it exploded below, its destruction range was far less as the upwards force of the explosion was partially countered by the speed of the falling bomb.

The Japanese developed a series of bombs for the purpose which sprayed incendiary pellets all over the place. Their attack was normally carried out from head on at a range of approximately 1,000 yards, with a mere 200-odd feet height advantage.

Both German and Japanese air-to-air bombing was relatively ineffective even against massed formations. If you wish to try it I suggest a 00 on the dice for a hit (not necessarily direct); a hit having been obtained, work it out as though it was a one second burst with a factor of 20 points.

Air-to-air rocketry was not so hot either. The Russians started it with their RS 82 but, judging by the fact that it was hardly, if ever, used during 1944-5 in the air-to-air role, it would not appear to have been very effective. The Germans used the 21 cm Nebelwerfer 42 for attacks on massed bomber formations. Fired from what looked like an outsize cocoa tin under the wings, it was time fuzed to explode 600-1,200 yards from the launch point. If it connected it was deadly, but it seldom did, its gravity drop being far too high for accurate aiming. The Germans ended the war using the R4M rocket which, being much faster than anything previously tried, was considerably more accurate. The normal mode of attack was to ripple fire the lot at a bomber from 600 yards or so. The dispersion gave a reasonable chance of hitting if enough went off at once.

Generally speaking, the use of the rocket as an air-to-air weapon was to enable the attacking aircraft to stand off at long range and take pot-shots. Going in with rockets to point blank range would rather defeat the object of the exercise and, bearing in mind that a single hit would usually bring down a heavy bomber, we can assess their effectiveness as follows: Range of all weapons 300 mm. Throw 00 or 01 for a hit for each pair of rockets fired. If salvo or ripple firing, add one for each extra rocket in the salvo, ie, 12 rockets = 11 per cent probability. With an 11 per cent probability, if 05 is thrown, it indicates two hits, and if 02, three hits. Calculate each hit as having a factor of 20. Nebelwerfers—00 per weapon to hit, fire factor of 30.

Chapter 7

Air-to-ground and ground-to-air

So far we have been mainly concerned with the fighter, both in its air superiority and its interceptor roles. Essentially, World War 2, as viewed by the major powers, was fought with the intention of eliminating the capacity of the enemy to resist so that his country could be occupied; the occupation being necessary to prevent his country from rearing up on end at some future date. This occupation would normally be carried out by the army, as only they could effectively subdue territory. This is not to say that air operations were only a side line, but they were, in fact, a means to an end, the means being to strike at surface targets, whether on land or at sea. This was carried out in many different ways, but the principle of cost effectiveness predominated. Whether the cost was calculated in economic terms or in lives could vary according to the task in hand, but the principle remained. For example, the cheapest way of immobilising an enemy tank was to put a handful of sugar in the fuel. As the enemy would be likely to object to strangers walking around his tanks with a barrow-load of sugar, it can be seen that some other means had to be found. The air force method would be to drop bombs on it, discharge rockets at it, or shoot at it with guns. In terms of aircraft lost, fuel and ammunition expended and the enormous cost of keeping aircraft and pilots fuelled and serviced, the economics of destroying tanks does not at times seem all that advantageous. On the other hand, the cost to the ground forces in terms of money, effort and lives would almost certainly be worse. Also, anything that speeds up the war is to be preferred. A quick battle is likely to be cheaper than a long drawn out battle. The best proof of this is the Blitzkrieg of 1940 when compared with the 1914-18 war.

Another illustration would be aircraft factories. Let us assume that the Weltkrieg Flugzeugwerke AG produces ten fighters per week. A 50-bomber raid is laid on with suitable escort, and loses eight bombers at £40,000 each plus two escort fighters at £8,000 each. Leaving aside the back-up services but bearing in mind the cost of training crewmen, the cost of mounting this raid would be in the region of one million pounds. The factory takes a fearful beating and does not recommence production for six weeks, returning to full production after 12 weeks. Ignoring the the cost of factory repair, replacement of essential manufacturing equipment, wastage of raw materials and effort, etc, the enemy air force has lost between 90 and 100 fighter aircraft. The value of this cannot be assessed as they have never been built, but if each fighter lost was able to take to the air and shoot down half a bomber each, the bomber force future losses would be at least 45. This is all obviously hypothetical but you can see how the principle works.

Air Battles in Miniature

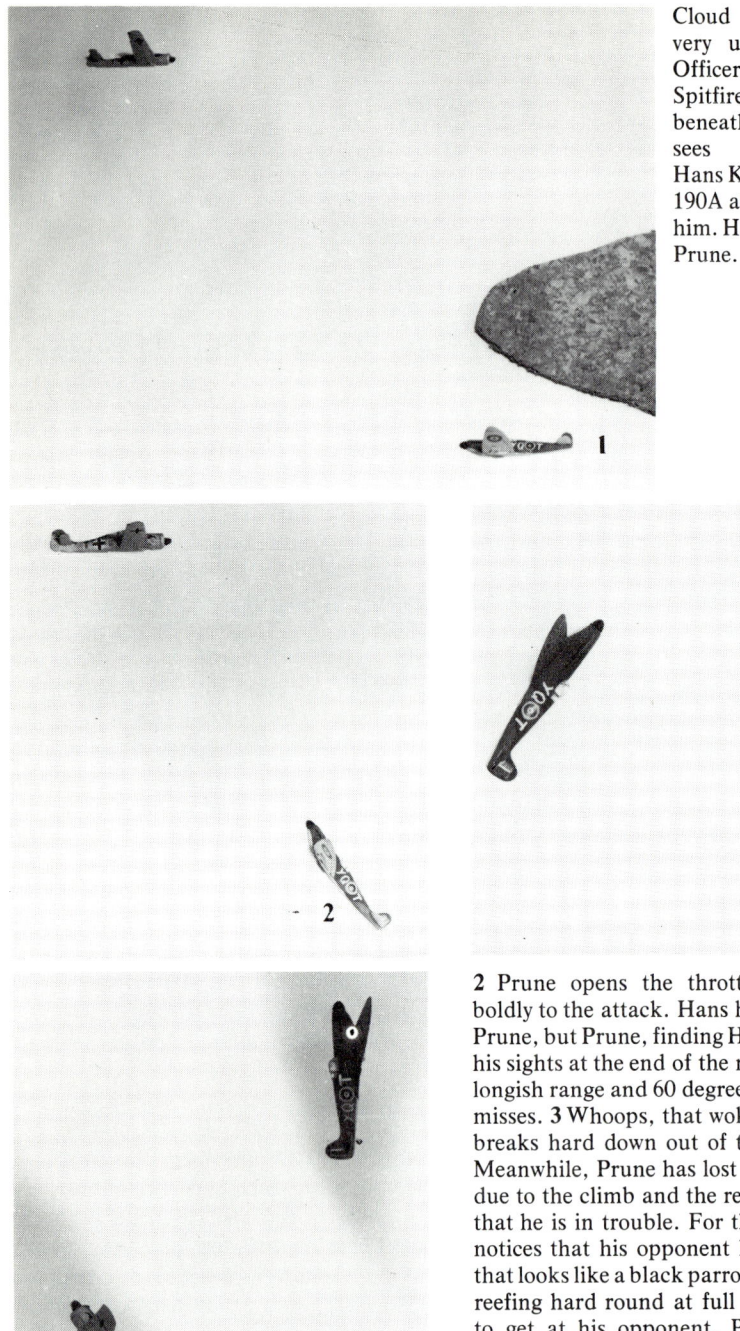

Cloud cover can be very useful. **1** Pilot Officer Prune in a Spitfire emerges from beneath a cloud and sees Oberleutnant Hans Kneise in an FW 190A about to fly over him. Hans fails to spot Prune.

2 Prune opens the throttle and climbs boldly to the attack. Hans has not yet seen Prune, but Prune, finding Hans squarely in his sights at the end of the move, lets fly at longish range and 60 degree deflection. He misses. **3** Whoops, that woke Hans up! He breaks hard down out of the line of fire. Meanwhile, Prune has lost so much speed due to the climb and the recoil of his guns that he is in trouble. For the first time he notices that his opponent has an insignia that looks like a black parrot. **4** Hans is now reefing hard round at full throttle, trying to get at his opponent. Prune for some unknown reason tries to reverse his turn and is now on the point of stalling.

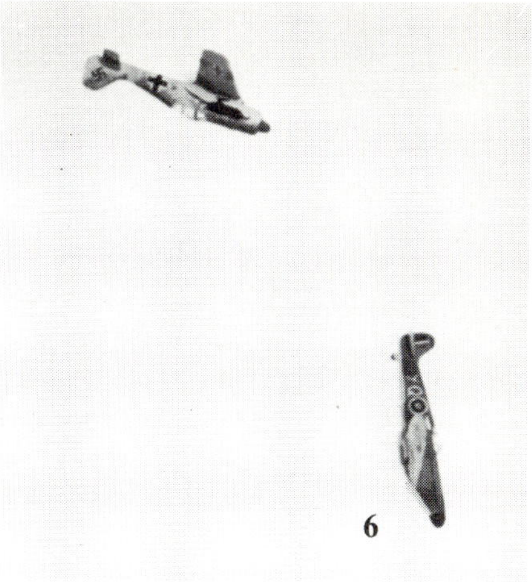

Air-to-ground and ground-to-air 63

5 Prune stalls, fortunately without spinning, Hans has throttled back but is still unable to bring his sights to bear. 6 Prune is falling vertically, his throttle wide open, trying desperately to regain flying speed. Hans is turning himself inside out to get on Prune's tail. 7 He succeeds as Prune recovers control, and knocks large lumps out of the Spitfire's engine as it heads for cloud. The engine, badly damaged, stops. 8 Good grief! Prune went into cloud with hardly any speed on the clock and a dead engine. Hans went in at high speed with his throttle wide open, and overshot, ending squarely in Prune's sights.

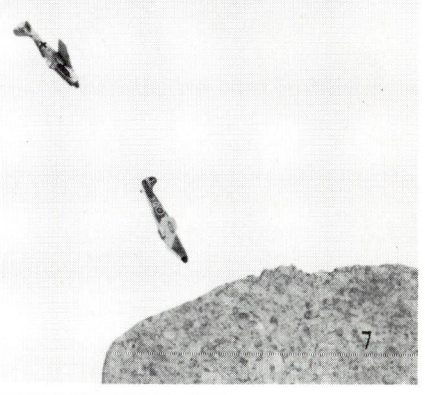

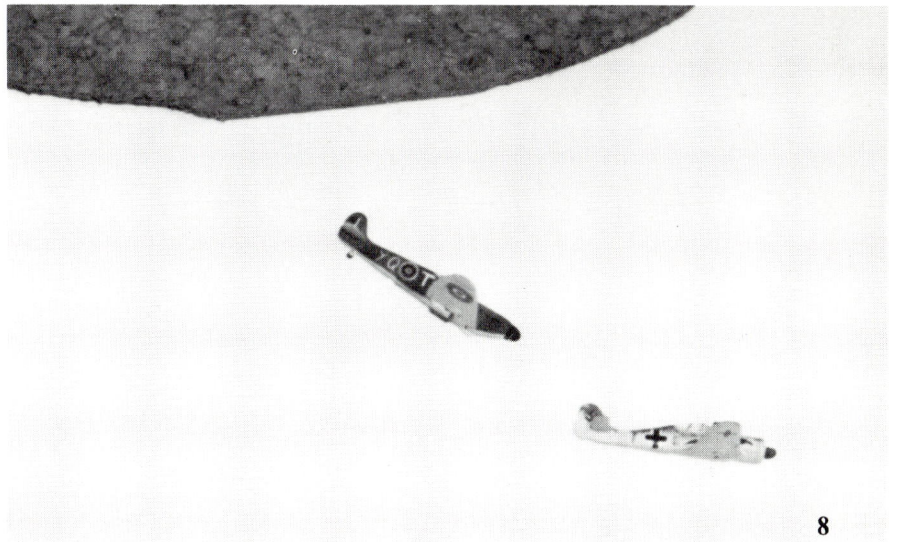

It would be true to say that the primary task of an air force is to carry out air-to-ground attack, whether this be long term strategic/economic, or the immediate destruction of a small enemy unit in the field. Viewed in this light, combat between fighter aircraft, regardless of the excitement and glamour, is subordinate to the needs of both the ground strike force and the surface forces. The prime function of the fighter force is to prevent the enemy strike force from carrying out its work, also to ensure that one's own strike force can operate unhindered, while the main purpose of the strike force is to get on with the war.

Air-to-surface attack takes many forms. The bomb is the primary weapon. It can be dropped from a great height by an aircraft in level flight. It can be dropped by an aircraft diving at a relatively shallow angle (glide-bombing). It can be dropped with great accuracy by an aircraft in a steep dive (dive-bombing). It can be dropped from a very low level with a delay fuze, often after bouncing from the surface, usually over water but not always (skip-bombing). Other weapons dropped were mines and torpedoes, also napalm. (Yes, its use dates back to 1944, and very nasty it was.) Rockets could be fired and, of course, guns.

To start with, let us look at bombs and bombing. The bomb was generally, though not always, streamlined, and could be one of several types. Incendiaries were generally small and designed to scatter during their fall, the purpose being to ignite any inflammable material they came into contact with and destroy by burning. Most of the early types used magnesium for the firing agent, but later in the war the RAF used phosphorous, which was much more effective. Incendiaries were most used against built-up areas, generally in conjunction with a blast-type bomb which chewed up the buildings sufficiently to give the incendiaries something to work on.

Next we have the high explosive, or blast-bomb, which varied in size from 25 lb to a full ten tons, the purpose of which was to produce a large explosion which flattened everything in the vicinity. This type sub-divides into armour-piercing, for use against tanks, ships, steel gun cupolas and fortified buildings, which featured a thick steel skin for penetration before exploding; the light case, or blast-bomb, in its most extreme form being little more than an overgrown dustbin full of explosive; and the general purpose, which was somewhere between the two, capable of more penetration than the light case and carrying more explosive, size for size, than the armour-piercing bomb.

The ultimate in free-fall conventional bombs was the Grand Slam, the ten-ton brainchild of Dr Barnes Wallis, the design of which was carried to such extreme lengths that the tail fins were offset, so that the bomb started spinning on its way down, the gyroscopic effect produced being sufficient to hold the bomb steady as it accelerated through the speed of sound, where otherwise turbulence would have deflected it slightly. When it struck, it bored deep into the ground before exploding, causing the most unbelievable crater. Being quite expensive, and needing a specially adapted Lancaster to carry it, it was only used on selected targets, such as U-boat pens, rail tunnels and, of course, the Bielefeld Viaduct which had survived months of attack. One near miss by a Grand Slam gouged out an enormous crater and a span or two collapsed gently, rattled into destruction by the camouflet, or shockwave effect.

Other types of specialised weapons were developed, such as the dams raid bomb, and the rocket bomb designed to crack the U-boat pens. Two tons in weight, this was a hard case (for penetration) bomb which fell normally to 5,000 feet, at which point a barometric fuze ignited a rocket, which boosted the speed on impact to over

Air-to-ground and ground-to-air 65

1,600 miles per hour. Over nine metres of reinforced concrete could be penetrated by this weapon, and a good description of what it was like to be on the receiving end can be found in *Iron Coffins*, by Werner Herbert, a U-boat commander who witnessed one such attack.

The V1 and V2 missiles have no place here as they were ground-to-ground weapons, and thus are more related to the artillery than the air force. Germany, however, did manage to produce and use a couple of 'smart' bombs; the Hs 293 and the Fritz X. Controlled by the parent aircraft by radio or with wire guidance, these achieved some notable successes, sinking the Italian battleship *Roma* and severely damaging with no less than three hits a British battleship, HMS *Warspite*. The wire-guided version of the Hs 293 was not appreciably limited by its wires as they gave a separation distance of over 17 miles. The restrictions were more those of the operator's eyesight and the visibility at the time of attack. There was a rumour prevalent in the Royal Navy at one time that interference from an electric razor was able to jam the radio-controlled Hs 293. I have been unable to confirm this but it seems unlikely. The USAAF used the 'Azon' guided bomb fairly extensively from 1944 onwards. Control on this, however, was limited to azimuth, elevation corrections could not be made.

The Japanese produced the ultimate 'smart' bomb in the Kamikaze, the most refined expression of which was the Ohka (Cherry Blossom). The pilot was sealed in, and dropped from the mother aircraft at a considerable distance from the target. He glided to within a few miles then started a rocket motor which propelled him over the last few minutes at speeds of up to 600 mph. Once the rocket motor was on, the Ohka was virtually uninterceptable. The damage caused by these, and indeed by all Kamikaze missions, was not as high as it might have been, due to the low standards of pilot training in the later stages of the war. If you wish to try your hand at a Kamikaze attack, you do not, repeat not, need a log-book.

Level bombing, in the early days of the war, meant exporting bombs in the general direction of the target. The vector sight in general use with the RAF gave tolerable results from 5,000 feet on a clear day at a speed of 120 mph, but a bombing run at 5,000 feet on a clear day at a speed of 120 mph made the bomber an easy target for the defences. In order to survive, the bomber had to fly higher, faster, and in weather far from perfect, all of which aggravated the problems of accurate bombing. To use the vector sight, one fed in the aircraft's speed (air speed, not speed over the ground), altitude—another source of error as the altimeter worked on barometric pressure and changing weather conditions could alter the reading by a considerable margin, and *estimated* wind speed and direction; another potential source of error. A 10 mph wind error would be worth nearly 150 yards on the ground from 15,000 feet. In addition the aircraft had to be flown exactly level in both longitudinal and lateral planes. An aircraft flying with its wings as little as five degrees out of level would be nearly a quarter of a mile off target from 15,000 feet. Add half a second delay in pressing the button and at 200 mph you have a further 50 yard error to add to all the others. Now imagine that your target is a rail bridge over a river, approximately 18 feet wide and 80 feet long which is barely visible from 12,000 feet. What chance have you of hitting it, even though in this instance it is not defended, so that you haven't got flak to distract your aim?

Later in the war, the gyro-stabilised vector sight was introduced which gave considerably better results. It had a small integral calculator which automatically handled some of the bombadier's problems. One little trick which could only be worked with the vector sight was the use of an offset aiming point. If a bomber force was trying to hit a not easily visible target, often at night, a distinctive point nearby, such as a lake would be chosen as an aiming point, which would be marked in the normal way by Pathfinders. Other Pathfinders would measure the wind in the target area and report to the raid controller, or master bomber, who averaged these reports, then combined the result with a vector between the aiming point and the target. This came out in the form of a 'false wind' setting which was broadcast to the remainder of the force, who fed the 'false wind' into their bombsights. They then aimed at the aiming point and (hopefully) hit the target.

The other type of sight used for level bombing was the tachometric type, typical of which were the American Norden and the German Lotfe. This was a much more sophisticated device than the vector sight and in clear visibility was considerably more accurate. Bomb details and altitude were fed into it at the start, then the sighting telescope was lined up on the target and held there. The movement of the bomb aimer in holding the target in the sight sent signals into a computer which automatically told the pilot of any course corrections. The actual release of the bombs was also carried out by the computer. The disadvantage was basically that you needed a minimum 20 second bomb-run, which at 200 mph meant that you had to visually identify the target a long way back which, in the cloudy skies of Western Europe or industrial areas full of haze, was not so easy. The Royal Air Force did use a tachometric sight, the Stabilising Automatic Bomb Sight, or SABS, but not in any great numbers, prefering the Mark 14 Vector Sight for general use.

Bombs in free fall accelerate at the rate of 32 feet per second per second until at the fast end of the scale the surface drag begins to build up. However, I suggest we ignore surface drag for the purposes of this exercise. Remember that we are playing to a four-second move time. On the fourth second (first move) after release the bomb would be falling at 128 feet per second (87 mph); on the eighth second (second move) 256 feet per second (175 mph); on the twelfth second (third move) 384 feet per second (262 mph); the sixteenth second (fourth move) 512 feet per second (349 mph); and so on. In addition the bomb does not fall vertically but in a parabola, its exact course being determined by the speed of the aircraft at the moment of release.

Some idea of the rate of fall is given in the following table, which is related to our four-second game move time, and centimetre game height scale.

Move	Time (seconds)	Speed of fall Feet per sec	mph	Distance of fall (approx) Feet	cm
1	4	128	(87)	256	(2)
2	8	256	(175)	1,000	(9)
3	12	384	(262)	2,300	(19)
4	16	512	(349)	4,100	(34)
5	20	640	(436)	6,300	(53)
6	24	768	(524)	9,000	(75)
7	28	896	(611)	12,250	(102)
8	32	1,024	(698)	16,000	(133)

From this it may be seen that a bomb dropped from 16,000 feet will take 32 seconds to fall and will strike the ground at nearly 700 mph, and that 133 centimetres represents the 16,000 feet on our playing surface, and no less than eight moves represent the 32 seconds falling time.

Air-to-ground and ground-to-air 67

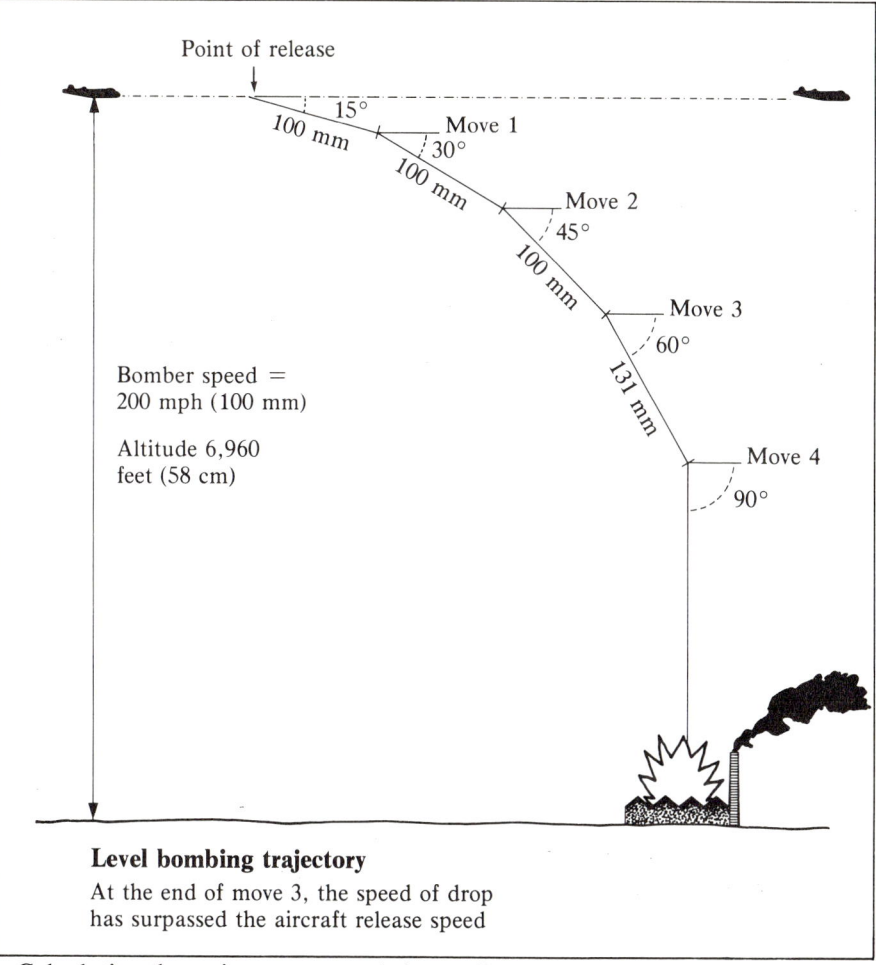

Level bombing trajectory
At the end of move 3, the speed of drop has surpassed the aircraft release speed

(Diagram labels: Point of release; 100 mm, 15°, Move 1; 100 mm, 30°, Move 2; 100 mm, 45°, Move 3; 131 mm, 60°, Move 4; 90°; Bomber speed = 200 mph (100 mm); Altitude 6,960 feet (58 cm))

Calculating the trajectory correctly is a tedious mathematical exercise, and for the purposes of the skirmish game I propose to use something much less demanding. When a bomb is released, it still retains the forward speed of the releasing aircraft, and continues travelling in a forward direction for quite some time. If we then say that on release, the bomb falls away from the aircraft at an angle of 15 degrees for the first move, 30 degrees for the second move, 45 degrees for the third move, 60 degrees for the fourth move and vertically thereafter, travelling a distance each move corresponding to the speed of the aircraft on release *until this speed is exceeded by the speed of fall in the table, when the speed of fall in the table is used.*

Ballistics whizz-kids, logicians and mathematicians can pick holes in this method, *but it works*! If you use a pin as a marker for the bombs as they fall you will end with a precise point of impact. Pin-point accuracy, in fact!

Having dealt with level bombing we next go on to glide-bombing. This subdivides into shallow glide-bombing, generally carried out at low level, 2,000 feet (17 cm on the playing surface) or less, at an angle of descent of up to 20 degrees; and steep glide-bombing, which technically takes place between the angles of 21 and 60 degrees, from a relatively high altitude, 8-9,000 feet being fairly typical. Glide-bombing is so called because it is carried out with the engines throttled back to minimise acceleration in the dive. It was necessary to adopt this method of attack where more accuracy was required than could be achieved with level bombing, but the attacking aircraft was neither stressed nor equipped for dive-bombing; alternatively, a dive-bomber squadron would use this method of attack rather than a true dive if their standard of training was below par.

As far as our game goes I feel that shallow glide-bombing, which is hardly ever attempted, could be dealt with in exactly the same way as level bombing, with the angle of fall for the bombs taken as an angle from the line of flight of the dropping aircraft. Steep glide-bombing can be dealt with under the heading of dive-bombing.

Dive-bombing is officially classified by the angle of dive, which is between 60 and 90 degrees. Accuracy is, of course, directly proportional to the release height and, to a lesser degree, the angle of dive. An unopposed attack by well-trained aircrew in good weather conditions could be very accurate indeed. A measure of opposition made quite a difference however, as a minimum aiming time of ten to 12 seconds (three moves) was generally required.

A fully trained and experienced squadron should be able to land all its bombs within a 25 yard radius circle in an unopposed attack, in ideal conditions, releasing the bombs from about 2,500 feet. This is marvellous for clobbering the local gasworks but not so hot if you are aiming at one solitary tank which, working to a true ground scale, would be little over one millimetre long. What we must do therefore is to postulate a point of aim, work out the error, and see if the miss is sufficiently small to hit (quotation from Irish Air Force handbook). For example, a target 40 yards long (one centimetre on the table) with the aiming point in the centre, when bombed with an inaccuracy of 20 yards (five millimetres on the table) would be hit at one end. However, let us return to the 25 yards radius circle, and assume that the chance of putting a bomb bang (no pun intended) on top of a tank in the centre would be 0.7639 per cent. The approximate area of a tank is 15 square yards, the area of the 25 yard circle is 1,963.5 square yards. In other words, about three quarters of a chance in every hundred. Not very high is it? To go to the other extreme, try adopting the following formula and bear in mind that this is an air game, and that should you wish to incorporate air power into land and sea battles this will be dealt with further on, encompassing the third dimension.

Release height of bomb ten centimetres, and angle of dive 90 degrees gives a 100 per cent probability of a hit on a target one centimetre long. That's right, you can make sure of hitting it, but you crash as well! Deduct one per cent for each extra centimetre of height above ten centimetres. Deduct one per cent for each degree that the aircraft varies from the vertical (for example, a 75 degree angle of dive will give a deduction of 15 per cent). If a dive-bomber is damaged during its attack, halve all probabilities.

The difference between the probability and the dice throw is the bombing error distance in millimetres. If the dive angle is 60 degrees or steeper, call it an undershoot, otherwise an overshoot.

For fighter-bombers the usual method of attack was to dive at between 45 and 60 degrees, line up the target in their gunsight, ease back on the stick, count three, then release. Calculate as for dive-bombers, then divide the result by four.

We have now covered bombing as far as it concerns the two-dimensional game, and will shortly deal with it again insofar as it affects a surface battle, together with

Air-to-ground and ground-to-air

its associated weaponry, rockets and torpedoes. The final item, before we leave two dimensions for three, is surface-to-air firing.

Anti-aircraft guns split into three categories; heavy, medium and light, the arbitrary divisions being 75 mm and upwards are heavy, 37 mm to 57 mm are medium and 20 mm are light. This must not be taken as gospel as some nations classified their guns differently.

A brief look at a few of the main types is called for, starting with the heavies. Possibly the best known anti-aircraft gun of World War 2 was the German 8.8 cm Flak 36, which fired a shell weighing slightly over 20 lb to an effective height of just over 26,000 feet. The effective height is defined as the highest altitude at which the target can be engaged for 40 seconds (ten game moves). The shell, of high explosive, was detonated at a preset height by a clockwork fuze, splitting into something like 1,500 fragments of differing sizes, which could inflict lethal damage to an aircraft out to a distance of 30 yards from the point of explosion and capable of making holes at 200 yards range. If fired to a height of 20,000 feet, the shells took something like 14 seconds to arrive. Generally, the 8.8 cm Flak 36 was deployed in four-gun batteries and fired in salvoes, although in the night defence of the Reich much larger batteries were commonly used. With visual tracking taking about ten seconds, fuzing and loading a further five (all heavy anti-aircraft guns were single shot jobs), and 14 seconds on the way up, the time from the initial tracking to the explosion of the shells was very nearly half a minute. Once the track was complete and the gun laid, the time between shells in barrage firing would be the five seconds to fuze and load, although this rate of fire could not be kept up for long. Aiming every salvo would give a rate of fire of two salvoes (eight shells) per minute. All this was fine in clear conditions but bad weather was another kettle of fish.

The *Würzburg* gun-laying radar was developed fairly early on in the war, but otherwise you were stuck with sound locators. As these were very indefinite as to direction, and only had a maximum effective range of 6,000 yards, this meant that aircraft noise took over 16 seconds to reach the instrument. A target moving at 250 mph would be nearly 2,000 yards past its calculated position.

The British equivalent of the 8.8 cm Flak 36 was the Ordnance QF (Quick Firing) 3.7-inch, which was capable of hurling a shell weighing 28½ lb to an effective height of 32,000 feet. The effect of its shell was nastier than the Flak 36 previously described, being larger, but the rate of fire was rather less, taking eight seconds to fuze and load.

America produced heavy anti-aircraft guns in fairly substantial numbers, typical of which would be the 90 mm M1, which used a shell of comparable size and weight to the 8.8 cm Flak 36 but with a considerably better effective height. Of very similar performance was the Russian 85 mm Model 1939. Italian and Japanese heavy anti-aircraft guns seem to have been generally inferior to those already described.

The most widely used gun in the medium bracket was the Swedish Bofors, which was in service with almost all the combatant nations. Not particularly outstanding in performance but very reliable, it could churn out shells weighing just under 2 lb at an average rate of 75 per minute, up to an effective height of 18,000 feet.

The German equivalent was the 3.7 cm Flak 36. With a practical rate of fire of 80 rounds per minute, a shell weight of only 1.4 lb and an effective height of nearly 16,000 feet, it was arguably inferior to the Bofors but not by much. Other nations produced medium anti-aircraft guns, but the ubiquity of the Bofors and its imitators was such that there does not appear to be much point in looking at them in detail.

We now come to the light anti-aircraft gun. The most commonly used of the German guns were the 20 mm Flak 30 and Flak 38, the Flak 38 having about double the rate of fire of the Flak 30, at around 200 rounds per minute. The effective ceiling was 6,600 feet and the weight of projectile 0.26 lb. A derivative of this was the quadruple-barrelled Flakvierling 38, which had four times the rate of fire, and which was frequently used in self-propelled mountings, or on trains. Most of the combatant nations, including Italy and Japan, made wide use of the Swiss 20 mm Oerlikon, either originals or rehashed slightly. The maximum effective ceiling was only 3,600 feet, but the rate of fire was in excess of 450 rounds per minute, with a shell weight almost identical to that of the Flak 38.

It is difficult to come by any comprehensive figures on the accuracy of anti-aircraft fire. From extensive reading I have concluded that heavy anti-aircraft guns, using predicted fire, used on average four thousand shells for each aircraft brought down. In other terms if a target was defended by 100 heavy guns, firing at an average rate of three rounds per minute, and the attack lasted for 90 minutes, the anticipated bomber losses would be six or seven, which in a heavy raid might be only one per cent or less of the raiding force, although three times this number would probably be damaged.

From this it can be seen that the heavy anti-aircraft gun, with its relatively slow rate of fire, can be omitted from the skirmish game altogether. Either the gun is fired at an average rate of three times per minute, or once every five moves, with fire effect not even taking place until the eighth move after the first sighting takes place, and then throwing *four* percentage dice, with a score of 0002 to hit; or, using just the pair of percentage dice, reducing the firing to once every 167 moves with a 00 score to hit, which is a waste of time, to say the least.

For the skirmish game, I would therefore suggest that heavy anti-aircraft fire is omitted altogether, and that medium and light anti-aircraft fire is considered on the same broad basis.

Medium and light guns were generally repeaters rather than one-shot jobs, and were normally fired at medium and low level aircraft, being aimed over open sights, frequently of the ring and bead type. At close range the 20 mm light flak in particular was deadly, and losses of up to half the attacking force could be experienced. The easy way out would be to take our air-to-air gunnery table, halve it for swivel mounted, and adjust the ranges suitably. This, however, will not do. As stated on page 57 the gunnery table gives better odds than real life, and this book is for air players, not flak gunners. Working on much the same principles however, we arrive at the facing table.

This can be used for all anti-aircraft fire. The effective range and points value of various weapons is:

Light machine-guns—150 mm—one point.
Heavy machine-guns—200 mm—two points.
Light ack-ack—300 mm—seven points.
Medium ack-ack—500 mm—20 points.

The application of points values and ranges is all as air-to-air gunnery. It should, however, be made clear that multi-barrelled weapons, like the quadruple flak mounting, count as one gun but with four times the firepower. Light flak can be fired three moves out of every four; medium flak, with its slower rate of fire, every other move.

Air-to-ground and ground-to-air 71

Ground-to-air gunnery

Speed of target	Deflection			Range in milimetres				
		50	100	150	200	300	400	500
Not above 150 mph	0-15°	15	15	12	06	00	00	00
	16-45°	05	05	04	02	00	00	00
	46-90°	02	02	01	00	00	00	00
151 to 300	0-15°	08	07	06	03	00	00	00
	16-45°	03	02	02	01	00	00	00
	46-90°	01	01	01	00	00	00	00
301 to 450	0-15°	04	04	03	01	00	00	00
	16-45°	01	01	01	00	00	00	00
	46-90°	00	00	00	00	00	00	00
Above 451 mph	0-15°	02	02	02	01	00	00	00
	16-45°	01	01	01	00	00	00	00
	46-90°	00	00	00	00	00	00	00

There are a considerable number of players who concentrate on land or sea battles but are aware that the air element should be represented in a realistic manner. Very few surface actions were fought during the war in which the air did not play an important, and sometimes decisive role, and it is to these actions that the remainder of this chapter is devoted.

Surface actions, by their very nature, are played in two dimensions. Our air game is also in two dimensions but unfortunately our two dimensions cover the vertical plane rather than the horizontal. While there is no earthly reason why a skip-bombing or torpedo attack cannot be made in the horizontal plane, it does rather pose problems if intercepting fighters are up, or if dive-bombing attacks occur at the same time. We could, of course, play the game in the horizontal plane using a method of recording the height changes, but it does make life rather complex, and the visual side of the air attack is completely ruined.

Then of course, an aircraft making a firing pass at a ground target has to be in a slight dive, and is constantly running out of altitude. Sooner or later it has to pull the nose back up, or risk making a large hole in the ground, and when it does this it is no longer able to fire. A tank on a reverse slope is a particularly difficult target, as the aircraft has to come in high enough to clear the top of the slope; stuff its nose down to get a shot in, then haul the stick back hard before it runs out of air space. If there was a line of tall trees a hundred yards or so behind the tank the problem would be compounded by having to clear them. Much the same applies to dive-bombers, where the release height is critical for accurate aiming. Can you imagine the problems of a dive attack in an Alpine valley? To a lesser degree the same thing applies in low level anti-shipping strikes, as the aircraft, having released its bomb at the last possible moment, has often to pull up to clear the ship.

Those of you who have read Bruce Quarrie's *Tank Battles in Miniature 3* will know what I am about to suggest, which is that we combine the horizontal and vertical planes. This is not as diabolical as it seems at first sight, although it is a little more complex than a straight air battle.

The land (or sea) game is set up as normal. The aircraft are set up as a normal air game, and when a ground target is observed, it is set up *in duplicate* at ground level, including any intervening terrain. The aircraft then attacks the duplicate target. Any anti-aircraft gun positions are set up in duplicate in the same way and fire according to the principles already laid down. For a full explanation see the following diagram A.

Air Battles in Miniature

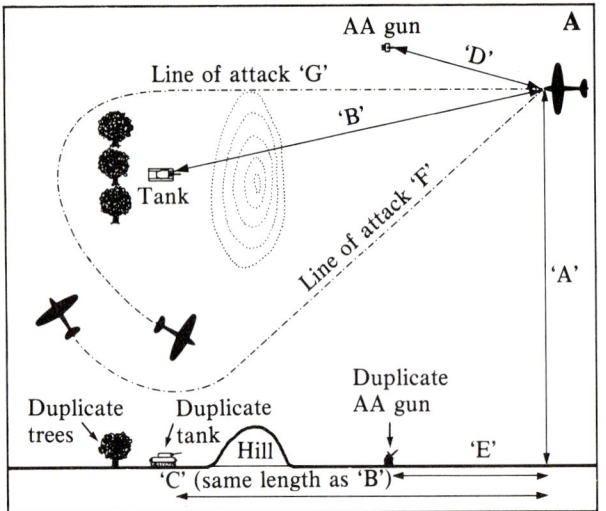

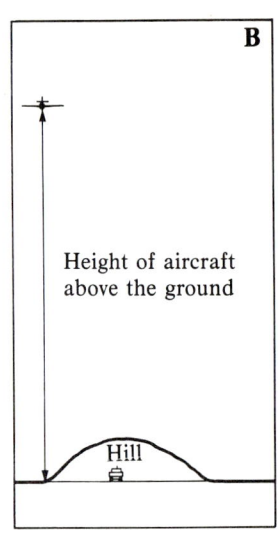

Enter the aircraft upstage right. This represents two things, its position in the sky in the vertical plane, and its position over the ground in the horizontal game. It observes the tank, sitting behind the hill, with a line of trees behind it, and announces its intention of attacking. Line A is extended from the spinner of the aircraft vertically downwards to give the point at ground level (vertical plane) that the aircraft is directly above. Line B is then measured from the tank to the spinner of the aircraft, this being the exact distance along the ground between the tank and the aircraft. This distance is then transferred to Line C, measured along the vertical plane ground level, and the duplicate tank set up at the end of it. This gives the exact three-dimensional relationship between the attacker and the target.

The hill and the trees are then set up on the same principle, making sure that their height is in direct proportion to the tank. As the tank is likely to be out of proportion to the ground scale, this is most important. Trees, for example, could easily be three or four times the height of the tank. Any anti-aircraft gun positions are set up on the same basis (Lines D and E) and the aircraft then carries out its attack. Two dotted lines are shown. Line F is the straightforward but easier line of attack, but will have the thick frontal armour of the tank to penetrate; Line G is a more indirect approach, going for the more vulnerable rear armour, but due to the trees getting in the way it has to be made in an excessively steep dive with a hard pull-out to avoid the hill. So much as a wing-tip touching it means a crash. If the trees weren't there the attack would be much easier. As it is, the pilot has to exercise a very fine judgement how close to press his attack before firing. A crash in this sort of situation is a distinct possibility. Diagram B, alongside diagram A, is simply to clarify things a bit.

You may say what a lot of fuss about air attacks, and I agree that if you are playing a game in a limited time, this method does hold up the surface action. On the other hand, anything less comprehensive than this is not air power at all, it's Snakes and Ladders! A suggestion or two. Dice for the move in which the air strikes arrive; they were notoriously unpunctual. If possible, adjust the air scale to match the ground scale you are using. If this is too awkward I suggest that you double the aircraft movement, ie, move one millimetre per mile an hour of speed. This works very well with the *Tank Battles* rules referred to earlier. Moreover, when an air

Air-to-ground and ground-to-air 73

strike arrives, play it through from start to finish if possible. If not, remember that there are, with a one minute time scale on the ground, no less than 15 moves in the air between ground moves.

Finally—cloud cover. This should not be ignored. Set clouds up in the horizontal plane. Observation is hindered if a cloud is in between the aircraft and the potential target. And to those of you who are fretting about getting on with the land battle when an air strike arrives, I can only say that as air power is so essential to the success of modern warfare, can you really settle for less? Where else are you going to get unskilled pilots writing off themselves and their aicraft?

Having established the *modus operandi*, it is time to have a look at the effects of the weaponry against ground targets.

The weapons were the gun, the bomb, the rocket and, by the US 9th Air Force in Normandy, napalm. The targets generally were buildings, armour, 'soft' vehicles and men. The performance of aircraft guns has been covered in depth previously, it remains to assess their performance against the various types of ground targets which we shall do later, together with rockets. What we really have to examine in depth for the land game is the effect of bombs.

I will not weary the reader with a long discourse on the merits and demerits of the various types of bomb, nor of the various ways in which a bomber could be loaded with an assorted mixture, to produce an almost infinitely variable range of effects on the ground, as this would be unacceptable to the average player who wants to get on with the game. Suffice it to say that there were almost as many types of bombs as there were types of bombers, and that the subject of explosives is itself an extremely complex subject, and they were constantly improved during the war.

To give you but one example, early in 1977 I had a long conversation with Dr Nic Büskens, the Dutch explosives expert, on the effect and extent of blast, during which he covered three serviettes with calculations, some of which I can still decipher. One kilogramme of trinitrotoluene (TNT to you) would, on detonation, produce a pressure of 184,000 atmospheres, (an atmosphere is about 14 lb per square inch at sea level), or 1,150 tons, at the centre of the explosion. Another way of expressing this is to say that one kilogramme of TNT, when exploded, turns into a white-hot gas which, when dissipated at normal atmospheric pressure, would occupy a space of no less than 20 cubic yards; or to give you something that you can more easily visualise, a gigantic sphere over 26 feet in diameter. And this is just the primary effect. Mind you, not all this sphere would constitute a total destruction zone, as this is simply the volume taken up by the primary expansion force of the explosion. The shock wave would extend much further out but the explosive killing zone would be only the inner zone of the sphere, as the effect of an explosion decays in direct proportion to the cube of the distance from the centre.

One further digression before we reach the nitty gritty. Some of you may have seen a reference to a more powerful explosive called Torpex which was created in 1942 from TNT by the addition of powdered aluminium. I hope that you wondered what on earth difference aluminium would make, because I am going to tell you anyway. When TNT is exploded, a black residue is left. For a long while it was though that this was plain soot, but chemical analysis revealed it to be a compound which needed just a soupcon of aluminium powder, two stirs and a wish as Granny used to say, to form an unstable, and therefore explosive compound all on its own, thus adding a bit extra to the original bang. But back to bombs.

For the purpose of the game we will assume that bombs are all alike in construction and also assume that about half the total weight is explosive, the rest being casing, fins and detonator. If blast were the only effect to be considered, some wonderfully precise calculations could be made. Unfortunately there are secondary effects, such as the air displaced by the gas formed in the explosion, the shock wave, which can be quite considerable, and large lumps of casing flying about, to say nothing of the fire potential of the enormous heat generated. The possibilities are almost unlimited, and far too complex to be incorporated here, so unfortunately we once again have to take a short cut. General Guilio Douhet, in his book *The Command of the Air*, postulated 100 kilograms (220.46 lb) of active material, to destroy an area 25 metres in radius (2,348 square yards), this giving a ratio of destruction of 10.65 square yards per pound of explosive, or a radius of destruction of 1.84 yards per pound. From this we can calculate theoretical radii of potential destruction. It is, however, also necessary to say that a bomb load is a bomb load, for if we try to introduce the scatter effect of the various types of bomb loads, together with the built-in delay which was possible, it would take a week to prepare for a game. We shall then, treat each bomb as a single missile and calculate the effect accordingly.

Bomb effect radius

Weight of bomb (lb)	Radius of effect (yards)	Scale radius (mm) 1 mm-4 yds	Weight of bomb (lb)	Radius of effect (yards)	Scale radius (mm) 1 mm-4 yds
50	9	2	3,000	71	18
100	13	3	3,300	75	19
250	21	5	4,000	82	21
440	27	7	4,400	86	22
500	29	7	4,500	87	22
550	31	8	6,000	101	25
660	33	8	8,000	116	29
706	35	9	10,000	130	33
880	39	10	12,000	143	36
1,000	41	10	12,800	147	37
1,100	43	11	13,200	150	38
1,760	55	14	14,000	154	39
2,000	58	15	17,600	173	43
2,200	61	15	22,000	193	48

Having arrived at the radius of effect, we need to consider the bombing error. Yes, I'm afraid that the earlier methods were purely for fun, to be used in an aerial game. If we are to take our land battle seriously, we need to establish an exact point of strike in the horizontal plane. The problems of aiming are, of course, exactly as laid down before.

We will deal with level bombing first, starting with an explanation of 50 per cent circular errors. The 50 per cent circular error is a measure of bombing accuracy. It is the radius, centered on the target, into which the nearest 50 per cent of individually aimed bombs will fall. This method of assessment was adopted in preference to taking a straight average distance, to discount inaccuracies caused by equipment malfunction for example, which could land a single bomb three miles away and thus distort the average beyond the reasonable. It is, then, the radius of a circle into which *half* the bomb loads will fall.

Under good conditions, the tachometric sight would give 50 per cent circular error results of 100 yards radius from 10,000 feet, and 450 yards from 20,000 feet. In poor visibility these results would be likely to increase by 2½ times. The vector sight would give 225 yards error from 10,000 feet in good conditions and results to

Air-to-ground and ground-to-air 75

compare with the tachometric sight from 20,000 feet. In poor conditions, however, it was rather the better of the two, and poor conditions were the case more often than not. The claim of the US 8th Air Force was that they could land a bomb in a pickle barrel from 20,000 feet. Given a pickle barrel 900 yards in diameter, I see no reason to disbelieve it. In fact, hardly a pickle barrel remained intact in the whole of Germany by the end of the war.

Anyway, let us be optimistic and work on the clear conditions results for the tachometric sight. A little horse-dealing gives us 50 per cent circular errors of 75 yards at 5,000 feet, 100 yards at 10,000 feet, 300 yards at 15,000 feet and 450 yards at 20,000 feet. Remembering that these are 50 per cent circular errors, we also have to cater for the bad 50 per cent, and to be charitable let us assume that the 50 per cent circular error is also the average error, which will give us circles into which all bombs will fall of 300 yards diameter for 5,000 feet, 400 yards diameter for 10,000 feet, 1,200 yards diameter for 15,000 feet and 1,800 yards diameter for 20,000 feet. We must further define the height bands as up to 5,000 feet, 5001 to 10,000 feet, 10,001 to 15,000 feet and 15,001 to 20,000 feet, and anyone bombing from higher than 20,000 feet is chicken. What we now need is a series of circular pieces of Perspex scribed to scale to give bombing result circles, as illustrated, which are not to scale. If you are using our standard ground scale of one millimetre to four yards the circle for the 20,000 feet bombing results will need to be 450 mm diameter, the 15,000 feet result circle will be 300 mm diameter, the 10,000 feet result circle 100 mm diameter and the 5,000 feet result circle a mere 75 mm diameter. If your ground scale differs from our standard you will need to adjust it.

The bombing result circles are made up of concentric circles which are segmented by straight lines at every 30 degrees. The concentric circles are in multiples of 50 yards radius for five and ten thousand feet, multiples of 150 yards radius for 15 and 20 thousand feet. The line 0-180 degrees represents the track of the bomber with 0 degrees as the direction of flight. Potential bomb strike points are at the centre, naturally, and wherever a straight line meets a circle. Each circle is numbered, starting from the centre, and each line is designated by its angle. Each strike point can therefore be identified thus 2/60 degrees, or 3/270 degrees. The actual strike point for each bomb load is determined by laying the circle for the relevant height with its centre-point in the exact centre of the target, with the 0-180 degree line laid along the direction of flight, then roll the dice to determine the exact point of impact. On the 5,000 feet circle a score of 00 or 01 is a direct hit on the centre point; on all other circles 00 only is a direct hit on the centre point. Otherwise the point of impact is laid down in the tables on page 77.

The method of using the bombing result circles is as follows:
1 Determine the height of bomb release.
2 Nominate the target.
3 Lay the bombing result circle on the target, with the centre point in the middle of the target, with the 0-180 degree line on the track of the bomber.
4 Roll the dice to determine the exact point of impact.

For example, if you are bombing from 10,000 feet, and your dice score is 54, the point of impact as shown in the chart is circle 3, line 60 degrees, which is the point where the circle 3 is intersected by line 60 degrees. If your bomb-load was 4,000 lbs, anything within 82 yards (21 mm to our ground scale) is at risk.

Dive-bombing is a little more difficult to assess, as it is a complex of relationships between height of release, angle of dive, speed over the ground (which in a vertical dive would be nil) and the strength of the opposition. Without a fantastically

Bombing result circles

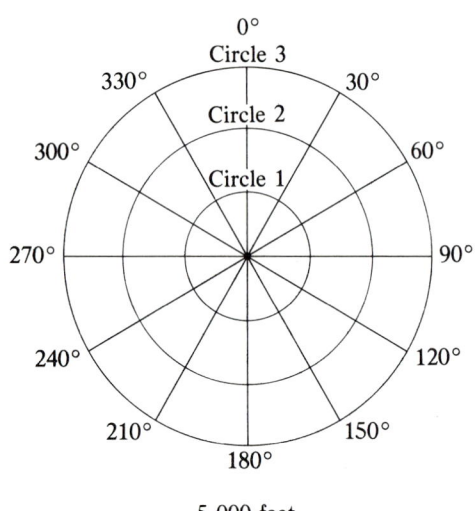

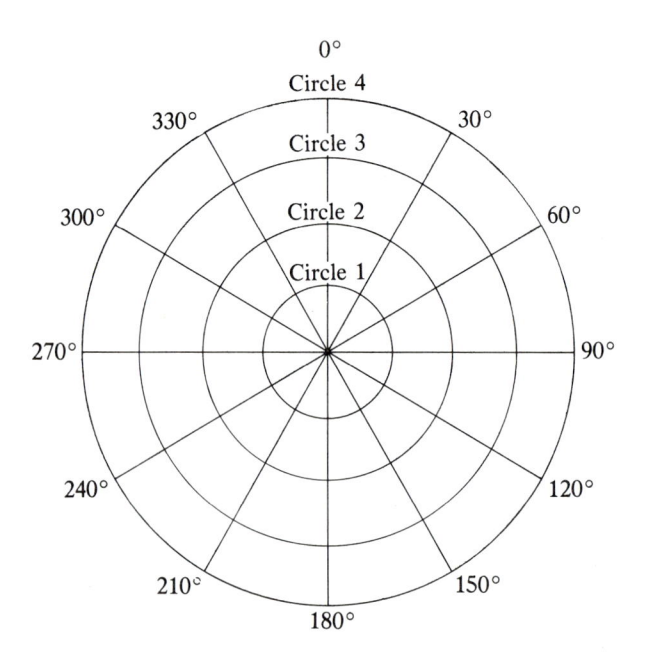

5,000 feet

10,000, 15,000 and 20,000 feet

Ground scale 1 mm = 4 yards

Air-to-ground and ground-to-air 77

5,000 feet line	Circle numbers 1	2	3	5,000 feet line	Circle numbers 1	2	3
0°	02-04	38-40	74-76	180°	20-22	56-58	87-89
30°	05-07	41-43	77-78	210°	23-25	59-61	90-91
60°	08-10	44-46	79-80	240°	26-28	62-64	92-93
90°	11-13	47-49	81-82	270°	29-31	65-67	94-95
120°	14-16	50-52	83-84	300°	32-34	68-70	96-97
150°	17-19	53-55	85-86	330°	35-37	71-73	98-99

10,000 and 15,000 feet line	Circle numbers 1	2	3	4	10,000 and 15,000 feet line	Circle numbers 1	2	3	4
0°	01-02	25-26	49-50	73-75	180°	13-14	37-38	61-62	87-89
30°	03-04	27-28	51-52	76-77	210°	15-16	39-40	63-64	90-91
60°	05-06	29-30	53-54	78-79	240°	17-18	41-42	65-66	92-93
90°	07-08	31-32	55-56	80-82	270°	19-20	43-44	67-68	94-95
120°	09-10	33-34	57-58	83-84	300°	21-22	45-46	69-70	96-97
150°	11-12	35-36	59-60	85-86	330°	23-24	47-48	71-72	98-99

20,000 feet line	Circle numbers 1	2	3	4	5	6
0°	01-02	17-18	33-34	49-50	65-66	81-83
30°	03	19	35	51	67	84
60°	04	20	36	52	68	85
90°	05-06	21-22	37-38	53-54	69-70	86-88
120°	07	23	39	55	71	89
150°	08	24	40	56	72	90
180°	09-10	25-26	41-42	57-58	73-74	91-93
210°	11	27	43	59	75	94
240°	12	28	44	60	76	95
270°	13-14	29-30	45-46	61-62	77-78	96-97
300°	15	31	47	63	79	98
330°	16	32	48	64	80	99

complex chart it is impossible to combine these factors in a satisfactory way, so I have developed a rule of thumb which seems to give satisfactory results in practice, based on the air game method but with a third dimension factor. Reading through, it may seem at first sight identical to the air game method, but it does differ in small essentials, so please read carefully.

Release height of bomb ten centimetres (1,200 feet) plus angle of dive 90 degrees gives a 50 per cent probability of a hit on the centre point of the target.

Deduct one per cent for each extra centimetre of height above ten centimetres.

Deduct one per cent for each degree that the aircraft's line of flight varies from the vertical. Minus percentages count as 00. Throw the dice. Should you not get a direct hit, the exact position of impact is determined by the number thrown. The left/right variation is determined by a second throw. Note both scores and adjust them as follows: The dice scores are related to distance as follows:

Height of drop	Adjust by
10 cm or less	÷4
11-20 cm	÷3
21-30 cm	÷2
31-40 cm	Nil
41-50 cm	× 2

Fig 8 Low-level bomb-sight

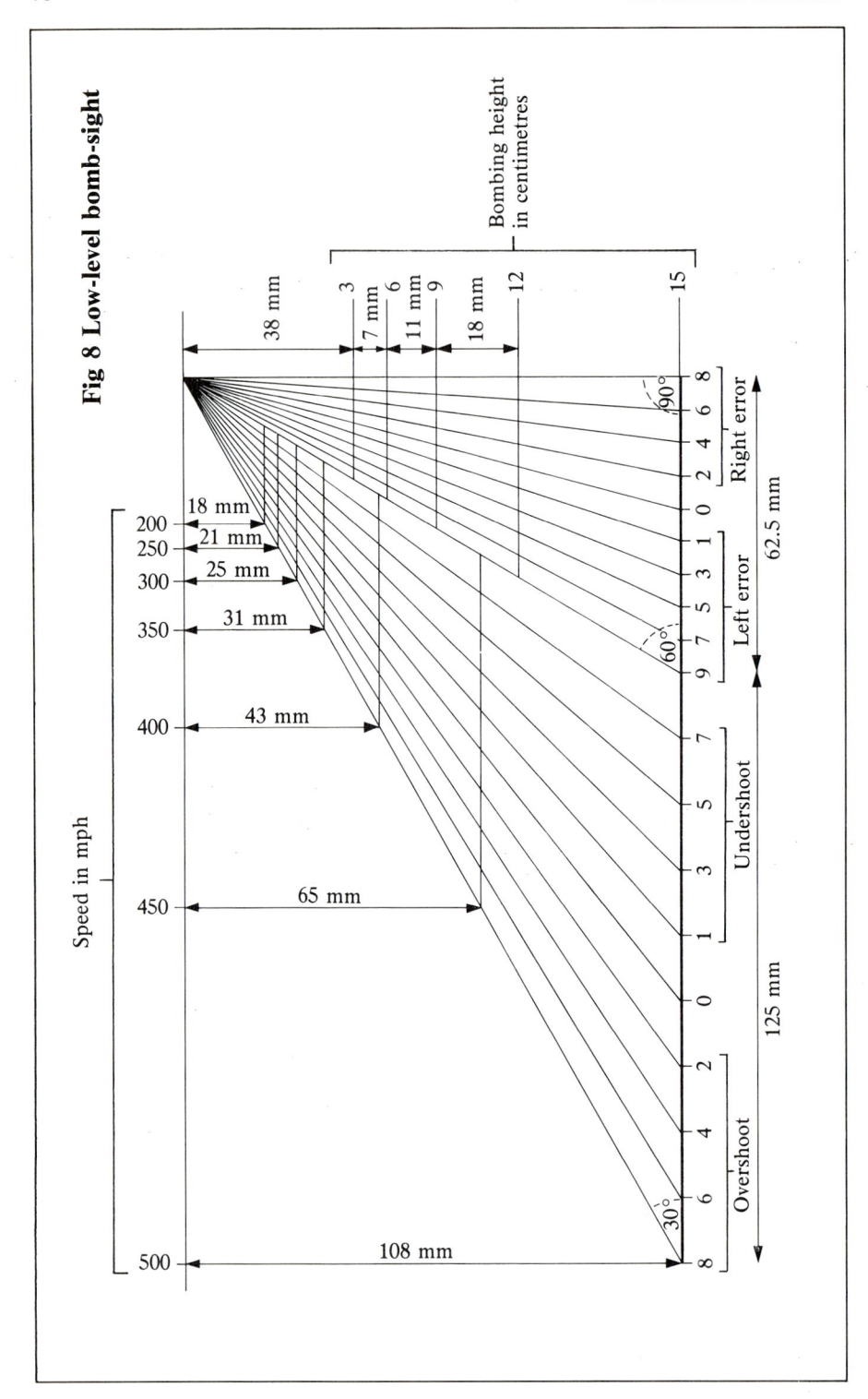

Air-to-ground and ground-to-air

For a score of 76 on a 29 cm height of drop, divide 76 by 2 = 38. This is the bombing error on the ground in millimetres. If the red dice score is higher than the black (which it would be in this case) it is an overshoot, otherwise an undershoot. With the same figure on each dice, toss up for it.

The second dice score is the left/right error, and the distance of the error is determined in the same way, the only difference being that if the number on the red die is larger it is a left hand error, otherwise it is a right hand error. If the same number is on both dice toss up once again. If the bombing aircraft is damaged during the attack, the probability of a direct hit is halved.

Fighter-bombers are a rather different kettle of fish, but if making steep attacks can be dealt with in exactly the same way as dive-bombers, although dividing their probability of a direct hit by two for each 50 mph or part thereof that their speed exceeds 200 mph. Thus a fighter-bomber attacking at low level at an angle of 45 degrees and a speed of 340 mph at the release point would first assess the probability as a dive-bomber, 50 per cent, minus nil for height, minus 45 per cent for angle = five per cent, halved for the first 50 mph excess speed = 2½ per cent, halved for the second part of 50 mph excess speed = 1¼ per cent, rounded off to the nearest one per cent probability. The bombing error is calculated in the same way as for dive-bombers, but with excess speeds in 50 mph bands replacing or adding to the height bands.

So far we have dealt with the fighter-bomber acting like a dive-bomber, making a relatively steep and sustained dive towards the target. They did, however, use another form of attack, which was to come in at low level and attack in a brief shallow dive (angle generally less than 20 degrees) or even release while flying level. Certain light and medium bombers also did this, and the method given here will do equally well for precision attacks on buildings, or skip-bombing of naval targets. It is based on the principle that lower and slower means greater accuracy, and variation due to angle of dive is immaterial. Bombs released in this way would tend to fall into a rectangular pattern rather than a circle, with the long side of the rectangle following the track of the aircraft. At this low level, overshoot and undershoot would be far more pronounced than lateral error; the Me 262 when used in this type of attack is recorded as having overshot by three quarters of a mile on occasion.

The method is to use the low-level bomb-sight as illustrated in Fig 8. This consists of two triangles, drawn to scale on squared paper. The left-hand triangle is concerned with the variable dropping error along the track of the aircraft. It is set up on a base line 125 mm long, with the sides going away at 30 and 120 degrees as illustrated. The base line is then divided into nine equal parts which are numbered as shown, and lines drawn from the divisions to the apex, which should, if it has been drawn accurately, be 108 mm above the base line. Horizontal lines are then drawn across the triangle at measured distances down from the top line. These horizontal lines are related to aircraft speed at the time of dropping.

The second triangle is a right-angled triangle with a base line 62.5 mm long. It fits exactly into the first triangle, the base is sub-divided and numbered as before, and the divisions connected to the apex. This second triangle is concerned with lateral bombing error, has horizontal lines which represent the height of bomb release in centimetres.

The method of using the bombsight is to roll the dice. The score on the red die is the longitudinal error, and the black score is the lateral error. For example, if the aircraft releases its bomb from eight centimetres up at a speed of 300 mph and the dice score is red seven, black two; taking the red score first, we follow the seven line up to where it is cut by the 400 mph line, then measure the distance to where the 400 mph line cuts the zero line, we find that we have a longitudinal overshoot of 22 mm. We then deal with the black score in a similar manner. Following the two line up to where it intersects the nine centimetre height line, we find an error to the right of a mere three millimetres. A dice score of 00 is, of course, a direct hit.

It may be thought that a lot of effort is being spent in establishing the exact point of impact of a bomb that misses. The purpose of this is that a land battle scenario is often quite crowded and things should (and did) get hit by accident on occasions. Also, the larger the bomb, the greater the radius of effect, and in certain cases a near miss is almost as effective as a direct hit.

Having spent a lot of time on bombing, we now turn to rocketry. Rockets were used by the Russians from the outset, but their effectiveness can be judged by the fact that the Germans made virtually no attempt to produce their own version, thus implying that they were not unduly impressed. They were used extensively by the British and to a fair degree by the Americans, although the American system was a bit of a lash-up, with a cluster of three bazooka tubes hung under each wing. As a weapon they were never very accurate, and their low velocity meant that they suffered severely from trajectory drop. Over a fifth of them didn't explode on impact anyway. They had to be released while the aircraft was flying straight, otherwise they would bind on the launching rails or in the tubes. In a nutshell, they were crude. They did, however, have one great advantage. Their ability to penetrate pill-boxes and tanks was excellent. They could be devastating on other targets too. An acquaintance of mine was flying Beaufighters over Burma in 1945. His brief was to destroy trains, but as the Japanese had only about 17 left at this point, there wasn't much chance of finding one. Returning to base after one such fruitless sortie, he encountered a mule train with a Japanese army escort, bringing supplies up to the front. Consoling himself with the thought that it was after all, an enemy 'train', he attacked from 50 feet with a full salvo of rockets plus four 20 mm cannon at 300 yards range. A split second later he found himself flying through what he described as a 'shower of mincemeat', his aircraft getting so badly splattered that he returned to base hotly pursued by every bluebottle in Burma.

Considerable controversy rages about the accuracy of the rocket as a weapon. After considerable study I have concluded that in the hands of an expert and experienced pilot they could be deadly, but the average pilot was unable to get results. A lot depended on range, and angle of attack. A hit would be easier to score from close range, but the flak, particularly on the Western Front, was so intense that it was fairly usual to let fly from 1,000 yards or more, while the nearer the attacking dive was to the vertical, the less gravity drop would affect the trajectory. At a range of 1,200 yards and attack angle of 45 degrees, the average error would be in the region of 50 yards, with a strong tendency to undershoot. This means that, under these conditions, all rockets would tend to fall in a circle of 200 yards diameter. As the amount of explosive carried was not large enough to warrant messing about with a blast radius circle, we must deal with these weapons on a hit or miss principle more akin to gunnery than to bombs. The following table gives the probability of scoring a hit on various types of target. Rockets could normally be fired in pairs, or in a salvo. Roll the dice for each *pair* of rockets fired at the target.

You will notice that no mention is made of bridges as targets. This is because rockets were ineffective against them.

Finally we have air-to-ground gunnery. This can be related directly to air-to-air

Air-to-ground rocketry

Targets	\multicolumn{6}{c}{Ranges in millimetres}
	50	100	150	200	300	400
Vehicles, pillboxes, gun positions	10	05	03	01	00	00
Massed infantry, trains, buildings	20	10	05	03	01	00
Large buildings	30	15	08	04	02	01
Infantry widely deployed	30	15	08	04	02	01
Ditto in cover or dug-in	30	15	08	04	02	01

gunnery for accuracy, as almost any target worth shooting at will be approximately aircraft size. All firing will obviously be at nil deflection; relative speed will be the speed of the firing aircraft. The one exception will be when shooting up deployed infantry, as here you are spraying a large area, so large in fact that a hit can be assumed for each second of firing.

Having covered ground attack accuracy in some detail, it is now time to assess the effect on various targets. Potential land targets are infantry, vehicles and structures. As it says in the Bible that the last shall be first, we will start with structures.

Structures sub-divide into categories, the first of which is light, consisting of barns, sheds and timber-framed buildings generally, also wooden bridges. Second is domestic and consists of brick and stone dwellings and the like. Third we have industrial: factories, hangars, steel or reinforced concrete framed office blocks, and brick or stone bridges or viaducts; and fourth is heavy, consisting of steel or reinforced concrete bridges and reinforced concrete fortifications such as pillboxes. Finally we have superheavy, which are very rare, consisting of U-boat pens, and the Atlantic Wall, Maginot and Siegfried Line gun emplacements, of which fortunately only the Atlantic Wall is likely to concern us to any degree.

Vehicles are categorised as soft (ie, unarmoured), trains, light armour and heavy armour.

Finally we have infantry, which divideds into mass and deployed. Mass infantry can be marching up to the front, in which case someone is an idiot for allowing it, or in lorries or trains; being caught in which could happen to anyone. Deployed infantry are invariably well spaced out and less vulnerable, and automatically fall flat when under air attack. They will normally be in three sub-divisions—in the open, under cover (woods or light buildings), or dug-in (trenches, foxholes, gun emplacements or domestic buildings).

The effects of bombs on these categories of target are as follows:

Structures

Light—Destroyed if the radius of effect so much as touches it. 50 per cent probability of catching fire.

Domestic—Destroyed if part of target lies within two thirds of the radius of effect. Otherwise damaged. Two lots of damage (assessed cumulatively) = destroyed.

Industrial—That part of the target which lies within half of the radius of effect is destroyed; the rest is damaged. Three lots of damage = destroyed. Bridges in this classification must receive a direct hit of 500 lb or more.

Heavy—Must receive a direct hit of 500 lb or more, bridges a direct hit of 1,000 lb or more.

Superheavy—Must receive a direct hit of 12,000 lb or more.

Vehicles
Soft vehicles—Destroyed if within the radius of effect; immobilised if within 1¼ times this radius and damaged at 1½ times the radius of effect.

Trains—If a direct hit is scored, total destruction will take place out to the radius of effect. In the event of a near miss, any part of the train within one quarter of the radius of effect will be destroyed, any part within half the radius will be damaged and immobilised, and any part within the remainder of the radius will be damaged.

Light armour—If within one quarter of the radius of effect will be destroyed, if within one half of the radius will be damaged and immobilised.

Heavy armour—Destroyed by a direct hit only up to 1,000 lb, below this no damage is caused by a near miss. Above this heavy armour is destroyed if it is within one eighth of the radius of effect, damaged and immobilised if within one quarter of the radius of effect.

Infantry
The probability of infantry in certain situations becoming casualties is shown in the following chart (RE = radius of effect).

Infantry casualty chart

	Impact point to ¼ RE	¼ RE to ½ RE	½ RE to ¾ RE	¾ RE to full RE
In the open	100%	50%	25%	12%
Under cover	80%	40%	20%	10%
Dug-in	50%	25%	12%	6%
Entrained or lorried	95%	45%	23%	12%

Each infantryman (or gunner) at risk is to be diced for; a score above those in the charts means that he survives intact. Anyone exactly under the point of impact is killed.

Rockets against the same target categories as bombs come out a little differently, particularly as near misses have no effect.

Structures
Light—Destroyed if hit.
Domestic—Destroyed by the cumulative effect of two hits.
Industrial—Destroyed by the cumulative effect of four hits.
Heavy—Pillboxes are destroyed if hit, all others are untouched.
Superheavy—No effect.

Vehicles
All categories are destroyed if hit. Only the part of a train where the hit occurs is destroyed.

Infantry
Casualties depend on the density of the troops. Deployed infantry cover a wide area; it is relatively easy to obtain a hit in this area but casualties tend to be either non-existent or light. A rough rule of thumb would be to say that infantry *en masse* have a 20 per cent chance of becoming casualties up to a maximum of 20 figures, and that deployed figures, regardless of cover, have a five per cent chance of becoming casualties up to a maximum of four.

We now come to the effect of guns against various targets. Guns would not generally be used against buildings of any description, being relatively ineffective.

Air-to-ground and ground-to-air 83

Some of the heavier anti-tank guns would penetrate solid buildings, but as they would normally be loaded with solid shot they would do little more than leave a couple of small neat holes. Shooting-up a building would, however, make the occupants keep their heads down.

Against vehicles they were far more effective. For each hit obtained on soft vehicles roll one dice to determine a strike factor, all as against aircraft. Four points destroy a soft vehicle, three points disable it.

Only cannon can damage armour and, as once penetration is achieved, the missile tends to ricochet around the inside, a hit is as good as a kill. 20 mm cannon were only effective against light armour, ie, less than 20 mm thick, although they were excellent for train-busting. The 40 mm cannon as carried by the Hurricane IID was effective against the armour of the PzKpfw III and IV, but nothing heavier. The German 37 mm weapon was effective against the Russian T-34 and KV-1 tanks from the sides and rear only, and the 75 mm gun would be likely to penetrate almost anything it came up against. The Russian 23 mm VYa cannon was effective against the side and rear armour of the PzKpfw III and IV, but nothing heavier, while the 37 mm N 37 cannon could even knock out the PzKpfw VI Tiger from the rear. Needless to say, the rear of a tank, where the thinnest armour lies, is the best place to go for.

Against infantry, the gun is a deadly weapon. Casualties for infantry *en masse* are assessed in the same way as rocket casualties; deployed infantry in the open have a ten per cent chance of being casualties up to a maximum of ten figures, and deployed infantry under cover have a five per cent chance of becoming casualties up to a maximum of five figures. They certainly keep their heads down when under fire, though.

Finally we have napalm to consider. The accuracy of drop is assessed as a bombing attack, and a typical missile would give a solid patch of flame, which nothing would survive, approximately 15 yards wide and 65 yards long (4 mm by 17 mm to our true ground scale), which would burn for several minutes, and make the surrounding area too hot to be tenable.

In order to make a successful attack on a ground target, it is first necessary to see it. This means that a) the target must be identifiable before the attacking run is commenced, and b) there must be no cloud or smoke between the attacker and the target during the attacking run.

Observation falls into two categories; air-to-ground and ground-to-air. We will deal with air-to-ground observation first. For this we again use the clock code, this time in the orthodox, ie, horizontal, plane, as diagram on next page.

The probability of observing a ground target per move is as follows: 12 o'clock to 3 o'clock and 12 o'clock to 9 o'clock—60 per cent; 3 o'clock to 4 o'clock and 8 o'clock to 9 o'clock—30 per cent; 4 o'clock to 5 o'clock and 7 o'clock to 8 o'clock—40 per cent; and 5 o'clock to 7 o'clock—20 per cent.

Further adjustments are made in the following order. Infantry targets count half these probabilities. For concealed targets divide the probability by four. If the target is a moving vehicle, add 20 per cent to the probability. If the attacking aircraft is being guided by a battlefield controller, or the target has been marked from the ground with a smoke marker, the probability increases by one and a half times. The first target seen will be attacked provided this conforms to the pilot's brief.

A word on the subject of battlefield control. This was extensively used by the Allies, and a reasonable approximation of the time it took to call up an air strike of

Air-to-ground observation

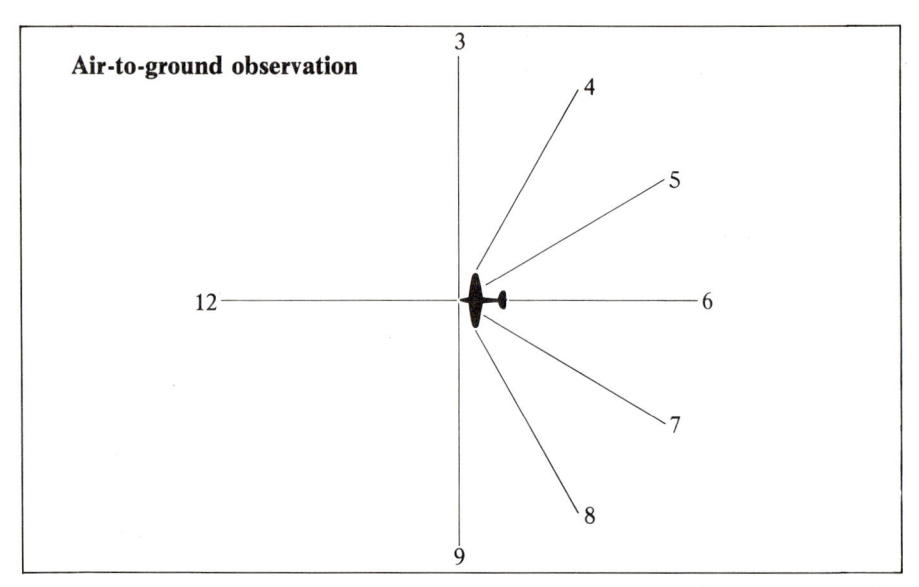

[Diagram showing an aircraft with clock-position directions 3, 4, 5, 6, 7, 8, 9, 12 indicated]

fighter-bombers would be about an hour and a quarter from first request to the attack commencing.

Ground-to-air observation is not quite as simple as you might think. On hearing an aircraft on a hot summer's day, the natural reaction is to glance up at it. If, however, a column of tanks is rattling and squealing its way past, or a spirited small arms action is being fought in the vicinity, this almost completely rules out hearing as a means of detection. In a probably confused battlefield situation, we are confined to visual detection, which operates as follows:

Units engaged specifically in the anti-aircraft role, unless under fire from the ground, will always spot either a single aircraft or a formation. Once they engage, they stay with it until it is a) shot down, b) out of range, or c) out of sight, when they will recommence their air search. Units engaged with either aircraft or surface units will be oblivious to all else until actually coming under fire from a different source. For all other units, probabilities are as the following table:

	Attacked from		
	Front	Flank	Rear
Engaged units	25	20	5
Units with enemy in sight	40	30	10
Unengaged infantry	100	90	75
Unengaged stationary vehicles	100	90	75
Unengaged moving vehicles	80	60	40

All units will become aware of a hostile aircraft when fired upon.

Having spent a lot of time on how to demolish the infantry, we now turn our attention to the requirements of the other lot, whose chosen life is to crawl around the surface of the globe in large metal contraptions called ships. Actually, we should be nice to the Navy. They were, after all, responsible for supplying we intrepid aviators with such essentials as petrol and, on overseas stations, with beer. But, of course, we are always nice to our own Navy, quite apart from those regrettable incidents of mistaken identity. And we didn't really *mean* any harm.

There is an awful lot of truth in the last couple of lines. The average flyer seemed

Air-to-ground and ground-to-air

at times to be totally incapable of differentiating between the *Bismarck* and the Isle of Wight ferry. This state of affairs was often reciprocated by naval gunners, who regarded anything that flew with all the affection that the Ancient Mariner displayed for the Albatross. Some of the errors made almost surpass belief. To quote a few examples, at the Battle of the Coral Sea in May 1942 the Japanese misidentified a refuelling ship and a destroyer as a carrier and a cruiser, while at Midway during the following month a formation of B-17s returned to base with the news that they had sunk a Japanese cruiser, which went down in 14 seconds! The sequel to this came a few days later when an irate American submarine commander demanded to know what the goddamned Air Force thought it was up to. He had been forced to crash dive.

We were not free of troubles on this side of the world either, HMS *Sheffield* surviving a torpedo attack by Swordfish after being mistaken for the *Bismarck* and a certain convoy from Alexandria to Malta which shot down all the Beaufighters sent to escort it. To go to the other extreme, during the Coral Sea battle, some rather confused Japanese gentlemen actually tried to land on the *Yorktown*. My general reaction is that even naval pilots had difficulty in telling their *Argus* from their elbow. Consequently I have been tempted to write a really diabolical couple of rules dealing with mistaken identity. But perhaps not.

Anyway, on to shipping strikes. Bombing can be done all as previously described, with the aiming point placed *where the bomb aimer thinks the ship will be by the time that the bombs reach it*. This is necessary as ships are moving, mobile targets, and in the open sea are very difficult to hit when they are taking evasive action at high speed.

One form of bombing generally used against shipping and not covered earlier is skip-bombing, known by the Luftwaffe as the 'Swedish Turnip' attack. The method was to come in low and fast, normally at less than 150 feet, and often below mast height, releasing the bombs by eye at close range, then pull up over the ship. The bombs were on a delay fuze to enable the dropping aircraft to clear the explosion, but this was distinctly hazardous for any aircraft following up, and it was not unknown for attacking aircraft to fly straight into the explosion of the bombs dropped by the preceding wave. Merchant ships were singled out for this form of attack for two reasons: 1) they were carrying supplies for land-based forces and/or industry; and 2) warships were much more heavily armed and armoured and consequently there wasn't much future in using this type of attack against them.

Travelling at 200 mph, or nearly 100 yards per second, the bombs were released between two and three hundred yards back, and bounced from the surface of the water (like playing ducks and drakes) to penetrate the side of the ship above the waterline. The ship had to be solid enough and tall enough to stop the bomb, which was easier said than done, as bombs dropped in this way have been known to bounce up to twice the height of the releasing aircraft, losing very little speed in the process; and being pursued by one of your own delayed action bombs, with the gunner apprehensively watching like a nervous slip fielder, was far from funny. In fact, the first air attack on a submarine during the war, made by a 233 Squadron Anson, resulted in the Anson ditching, having been damaged by shrapnel when its bombs bounced and exploded in the air. The submarine was undamaged; as it was one of ours this was just as well.

I have been unable to find figures relating to the effectiveness of skip-bombing; extensive reading and the application of common sense would appear to give a 25 per cent probability of scoring a hit by this method. As ships were generally well spaced out the chance of hitting a ship other than the target would be nil.

The next step is to examine the effects of bombs on ships. For this we will use the standard naval abbreviations which are: BB—battleships; CC—battle cruisers; CV—carriers; CVE—escort carriers; CA—heavy cruisers (8-inch guns); CL—light cruisers (6-inch guns and under); DD—destroyers; DDE—destroyer escorts (frigates, sloops, corvettes, etc); SS—submarines; MV—transports; T—tankers.

The other thing we need to do is to relate the ground scale of whatever naval rules we are using to our air strikes. My preference for naval action is *Naval Warfare 1939-45*, by Keith Robinson, published by Leicester Micromodels, which uses a scale of 1:20,000 or approximately one mm = 22 yards. As our air scale is one mm = four yards, we have problems, as the naval sea scale is nearly six times greater than our air scale. The easy way out, of course, is to play the naval end to the air strike scale for the duration of the attack. The principle of halving the speed in mph to give the move distance can equally be applied to the ships, and the ships can use turning circles six times larger than normal. This also has the advantage that the ship models used are not too far off being correct scale size, and our bombing 'near miss' data needs no alteration.

The effect of bombs on ships is difficult to rationalise, as the different categories of ship present greater or lesser degrees of vulnerability. A brief examination gives us the following assessments.

BB and CC tend to be very heavily armoured and consequently much less vulnerable than other ships. They are also virtually immune to damage from near misses. As the velocity of bombs is quite a bit less than that of shells, the penetrative ability is less. The so-called 'pocket battleship' should be classed as CC.

CVs are split into two main types; British and others. British carriers featured an armoured flight deck which acted as a shield between missiles and the very vulnerable aircraft with their highly inflammable loads. The deck armour could not, however, be made up to a really effective thickness without making the ships unacceptably top-heavy and, believe me, they were fairly top-heavy as it was. Carriers of other nations also generally had an armoured deck, protecting the vitals of the ship, but this was at hangar floor level, the flight deck being made of six-inch thick hardwood. This form of construction meant that bombs falling on the flight deck would almost always end up in the hangar, with consequent devastation. British carriers were less vulnerable to bombs than those of other nations, but more vulnerable to hits below the waterline, where flooding, unless compensated rapidly by counter-flooding, was liable to capsize them due to the greater top-weight. The carriers of other nations tended to be lost to uncontrollable fires making the ships untenable.

A fair amount has been written about carriers caught with aircraft on deck being more vulnerable. I cannot entirely accept this being so; the wooden flight deck being so easily penetrable it would seem more logical to postulate aircraft *on board* as an extra vulnerability factor. The main difference between armoured flight deck carriers and others was their resistance to relatively low velocity impact. For example, an American carrier on the receiving end of a Kamikaze attack would usually need a refit; on a British carrier it was 'all hands to the brooms', then business as usual. HMS *Formidable* actually took two unwelcome guests on board in one day without sustaining serious damage. Generally though, carriers, with their loads of bombs, torpedoes, aviation spirit, etc, were very vulnerable. They also made big targets.

Air-to-ground and ground-to-air

CVEs had all the main disadvantages of CVs but were generally unarmoured. They were often converted merchantmen, and carried a mere handful of aircraft, often permanently on deck.

CAs and CLs were scaled down battleships, with smaller guns and thinner armour, but generally faster.

DDs and DDEs were unarmoured, fairly mobile, and difficult to hit. They were, however, vulnerable to near misses and sufficient damage could be done in this way to sink them.

SSs were a waste of time attacking with conventional bombs except in very exceptional circumstances.

MVs, slow and unarmoured, were extremely vulnerable.

Ts, with their inflammable loads, were even worse.

Ship values can be arbitrarily assessed at one point per ton displacement, plus 20 per cent for BBs, CCs, and CAs, ten per cent for CVs and CLs; alternatively, refer to the naval rules previously mentioned.

Assume that each 250 lb weight of bomb will penetrate one inch of armour, for example a 1,000 lb bomb will penetrate a four-inch thickness of armour. Allow one damage point for each 1 lb weight of bomb if penetrating, and one quarter of a damage point if penetration is not achieved, or on DDs downwards; for a near miss, ie, a bomb 50 yards away or less, also one quarter of a damage point. Allow one and a half damage points per pound weight of bomb on CVs and CVEs to simulate their greater vulnerability, and two points per pound weight against tankers and ammunition ships. The thickness of deck armour varies widely; as a rule of thumb allow a quarter inch thickness for each 1,000 tons displacement to a maximum of ten inches on BBs, CCs and CLs, and two inches total thickness on British CVs only.

Torpedoes are the next anti-shipping weapon to be considered. These varied greatly in performance. At one extreme was the ineffective American 'fish' which had to be dropped in level flight from a height not exceeding 100 feet and a speed of no more than 125 mph. Very much a fragile 'this way up' sort of weapon, it also had a habit of not exploding on the rare occasions when it hit. At the other extreme was the Japanese Type 91 'Long Lance', both faster and longer ranged, which could be dropped from a 30 degree angle of dive from a height of 300 feet and at a speed of about 265 mph. British, German and Italian 'fish' were all somewhere between these extremes.

Tactics varied between nations. The Americans, handicapped by their poor quality kipper, attacked straight, low, and slow. The Japanese tendency was to attack like the clappers in a shallow dive, while the standard British attack was a steep turning dive at low speed, the speed being dictated by the antiquated biplanes used rather than any weakness in the torpedo. Twin-engined aircraft such as the Sparviero, Beaufort or Heinkel would generally attack at low level with a long weaving approach. Typical release ranges were one to three thousand yards which, with a typical speed of 40 mph, gave running times of between 50 seconds and two and a half minutes; plenty of time for a ship to evade. Pilots could, and often did, go in closer, but it really depended on the strength of the opposition. Nothing is so distracting to the aim as tracer close to the left earhole. And of course if you got in too close, the initial plunge of the fish was likely to take it clear under the ship anyway. Records show that somewhere between five and ten per cent of torpedo attacks scored hits, and that something approaching one in five hits didn't explode.

What I would suggest is drop your weapon and actually plot its track through the water, move by move, using air game length moves for both ships and torpedoes, *with only target ships permitted to take evasive action*. Could be fun in a large convoy. Give your torpedoes a range of 4,000 yards running (100 centimetres on the table) and see if you can't a) trap someone with simultaneous attacks from different directions, or b) sink someone by accident. For torpedo effect, see the Naval warfare rules; alternatively use the following formula. Roll the percentage dice plus one normal six spot die. Multiply the score on the percentage dice by 1,000, then divide by the number on the normal die. This is the points value of damage done. The choice of method is according to taste.

Before we leave the subject of torpedoes, mention must be made of an accoustic homer, which was lobbed into the wake of a departing U-boat. It was an antisubmarine weapon only, and is hardly worthwhile introducing into a game as so few were actually used.

Two other main weapons were used against shipping; the rocket and the depth charge. A salvo of 60 lb rockets was said to be the equivalent of a destroyer's broadside. Certainly it would be enough to severely damage the average unarmoured ship. I would suggest that ship targets be treated as large buildings for assessing hits, and each hit be treated as a four-inch shell under the naval rules.

Depth charges were, of course, anti-submarine weapons. Relatively ineffective until 1942, they then came into their own as the aircraft proved its ability to contain the undersea menace. It is an interesting fact that no less than 45 per cent of U-boat losses were due to air attack, 353 boats in all, compared with 31 per cent sunk by unsupported surface ships. A further six per cent were sunk by combined surface ship and air attack.

If you wish to mount an anti-submarine game, it should be borne in mind that one sighting was made in every 400 hours' flying time. In operation 'Enclose', for example, which was designed to clobber U-boats crossing the Bay of Biscay, the record shows that 41 U-boats crossed the Bay during this period, 26 sightings were made, leading to 15 attacks and only one sinking. Have I said enough to dissuade you? If not, read it all up in *Aircraft versus Submarine*, by Alfred Price; the definitive work on this side of the war. When reading German accounts, remember that they are in translation, and often refer to depth charges as bombs, the reason being that the German word is 'wasserbomben', or 'wabos' in the vernacular.

Ships, of course, carried defensive armament, and this tended to vary between ships of the same class. Again, we need a reasonable approximation to cover this, otherwise doing the anti-aircraft fire from even quite a small naval force would take so long as to become counter-productive.

Bearing this in mind, we must assess a firepower screening value for each ship based on the following assumptions.

1 Basis to be the number of heavy AA guns carried. This will include dual-purpose secondary armament provided that the secondary armament is not being used to engage surface target at the same time.

2 Light AA *should* be in proportion to heavy AA.

3 Heavy AA can cover a hemispherical volume about the ship of 4,000 yards' radius.

4 The hemispherical screen is split into two zones, one either side of the ship, to cater for the port/starboard gun mountings.

5 The screen should be further divided into four zones, 0 to 1,000 yards (Range

Air-to-ground and ground-to-air					89

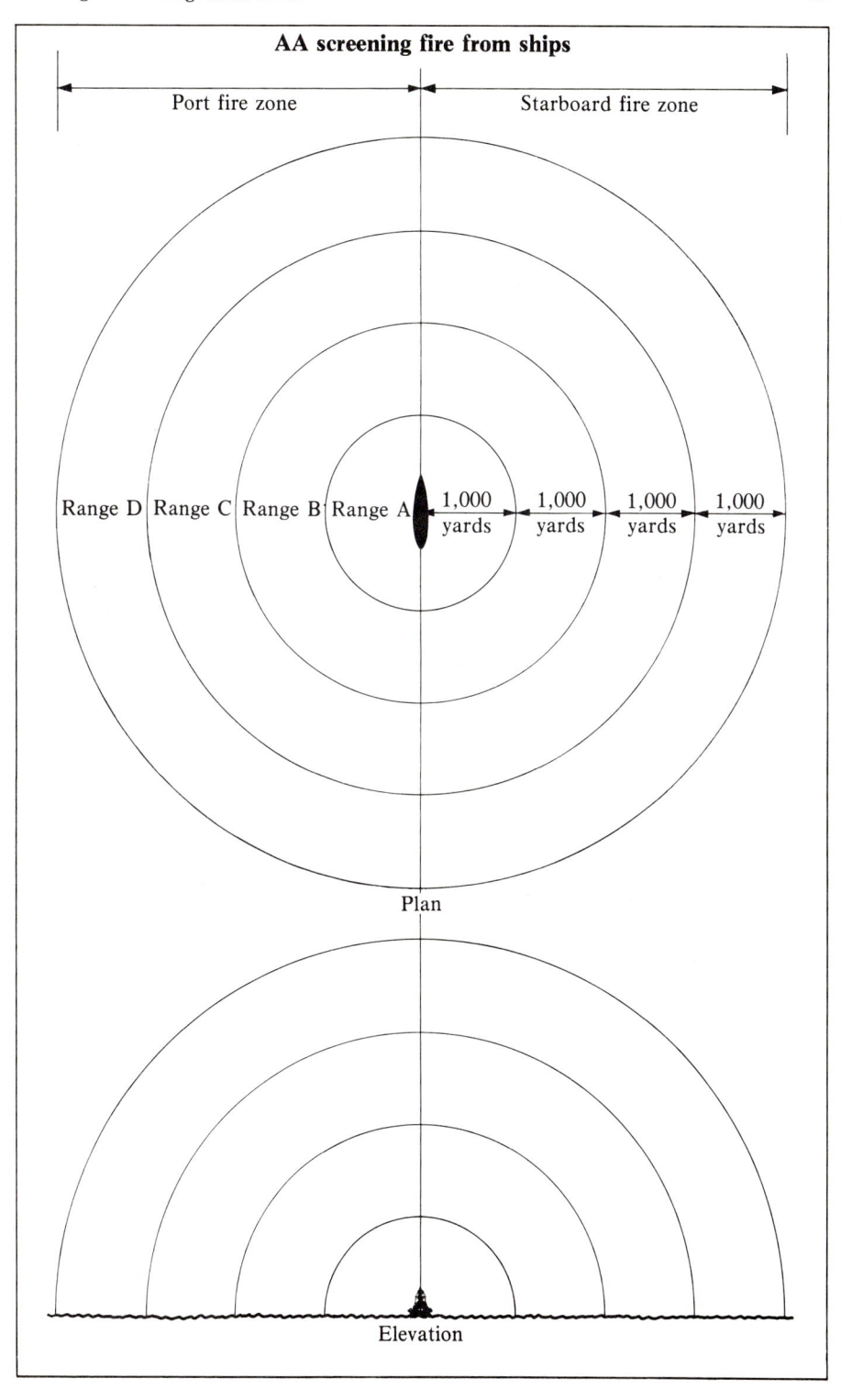

AA screening fire from ships

Port fire zone | Starboard fire zone

Range D | Range C | Range B | Range A | 1,000 yards | 1,000 yards | 1,000 yards | 1,000 yards

Plan

Elevation

A), 1,001 to 2,000 yards (Range B), 2,001 to 3,000 yards (Range C), and 3,001 to 4,000 yards (Range D). Range A should be the most deadly, diminishing in effect until the outer limit of the screen is reached. These zones and ranges are shown in the diagram of AA screening fire from ships.

These assumptions having been accepted, we can postulate a screening value, which is also the percentage probability of shooting down an aircraft, of the number of heavy AA guns carried by each ship. This applies to Range A only, but it also applies to both port and starboard firing zones. Thus a ship with eight heavy AA guns could engage single aircraft on both port and starboard at the same time, with an eight per cent probability against each. Probabilities are halved as the range moves outwards, thus an eight per cent chance in Range A becomes a four per cent chance in Range B, two per cent in Range C and one per cent in Range D.

It is, of course, possible for the AA screen to engage more than one target at the same time. If it is desired to engage three separate targets in a four per cent range, separate firing can be done at each of them at one per cent. Fractions are always rounded down in these instances. Once 00 is reached it becomes impossible to split fire any further. For example, with two potential targets in one per cent probability area it is possible to round down and have two shots at 00. With three potential targets in the same area only two can be engaged, as otherwise the AA effect will be increased. If it is advisable to engage two separate targets at different ranges, this can be done as follows: One target, if fired at alone would be worth, say, four per cent, the other target if fired at alone would be worth, say, two per cent. There are two targets therefore halve the probabilities down to two per cent and one per cent and have a go at both. It is, of course, necessary to state which targets are being fired at by what, before any firing is carried out.

If you are not a dyed in the wool naval player, I have prepared a list of screening values for a typical selection of ships, which is intended to be a rough guide only.

	Screen ranges			
	A	B	C	D
Royal Navy				
BB and CC Pre-war	08	04	02	01
Modern	16	08	04	02
CA	08	04	02	01
CL	08	04	02	01
CV	16	08	04	02
CVE	02	01	00	00
DD	06	03	01	00
DDE	04	02	01	00
MV, SS	01	00	00	00
US Navy				
BB Pre-war	08	04	02	01
Modern	20	10	05	02
CA	12	06	03	01
CL	08	04	02	01
CV	08	04	02	01
All others as Royal Navy				
Kriegsmarine				
BB	16	08	04	02
CC	14	07	03	02

	Screen ranges			
	A	B	C	D
Kriegsmarine continued				
Pocket BB	06	03	01	00
CA	12	06	03	01
CL	06	03	01	00
DD	04	02	01	00
MV, SS	01	00	00	00
Regia Marina				
BB Pre-war	08	04	02	01
Modern	12	06	03	01
CA	12	06	03	01
CL	06	03	01	00
DD	04	02	01	00
MV, SS	01	00	00	00
Imperial Japanese Navy				
BB Pre-war	08	04	02	01
Modern	12	06	03	01
CA	08	04	02	01
CL	08	04	02	01
CV	12	06	03	01
DD	06	03	01	00
All others as Royal Navy				

Chapter 8

The table-top game

Having waded through this far, you should be just about ready to try your hand. You will, of course, need a certain amount of equipment, quite apart from model aircraft and a surface on which to play. A tape measure graduated in either inches and tenths or centimetres and millimetres is a must, as is a pair of percentage dice. In case you are not familiar with these, let me describe them. They are 20-sided and are numbered 0 to 9; each number appearing twice. One die should be red and the other black. The pair of them are thrown simultaneously; the number uppermost on the red die constitutes the tens, the number on the black die is units. Thus a seven on the red die and a three on the black comes out as 73 (per cent). If you need a percentage probability of 16, then a die score of 16 *or less* is sufficient.

In addition you will need a turning device (described in Chapter 4) and a relative speed calculation diagram (also described in Chapter 4). Finally you will need an order sheet for recording height and speed; also writing orders for movement, etc. In order to simplify this I have prepared a heap of aircraft performance data for use in the game. As I have previously mentioned, sources tend to differ. If anything here does not square with your own source of reference, I would suggest that you assume that you are lucky (unlucky) enough to have a slightly faster (slower), or whatever it is, version. These things did happen.

The data given has been assembled to give a representative selection of aircraft types used by the main combatant powers, and is divided into sections according to function. As it would have been impractical to include every possible aircraft, your own favourite may not appear—the Halifax, for example. However, it should be easy enough for you to prepare your own chart for such aircraft. To save space, the information has been coded under the the following headings:

General 1 Year in which the aircraft entered combat in the service of the country of manufacture. It should be noted that many aircraft, mainly American, were used on active service by other nations earlier than the year shown. Also many German, Italian, Russian and Japanese aircraft saw combat in Spain and China well before the year shown. As the USA and Japan entered the global conflict on December 6 1941, for the sake of 3½ weeks I have listed 1942 as the year in which American and Japanese aircraft entered the war. A few major types have more than one sub-type listed. Where only one sub-type of aircraft is shown, this is the sub-type considered to be most representative of the type. For example, the Boeing B-17G Fortress has been taken as the representative sub-type although the year is stated as 1942 when the correct sub-type for this year would be the B-17D Fortress.

General 2 Normal loaded weight in pounds. This is the normal take-off weight in all cases with the exception of fighter-bombers and dive-bombers where, in order to compensate not only for the increased weight but also increased drag, I have taken the maximum bomb loads carried and added this to the normal take-off weight. The performance penalty thus arrived at is applied regardless of the actual load carried. Once, however, the load is shed the fighter-bomber will revert to its original fighter performance. This is not quite correct, of course, as the weight and drag of bomb-shackles, rocket rails or tubes,etc, remain, but a line has to be drawn somewhere. Fighters carrying air-to-air rockets, eg, the Bf 109-G6 carrying two 21 cm rockets, are also assessed as fighter-bombers for our purposes. In two cases I have departed from the truth. The Hurricane IV was the main rocket-carrying variant; the P-38L Lightning was the first rocket-equipped variant, not the P-38J, but I have incorrectly armed both the Hurricane IIc and P-38J with these weapons rather than introduce two new and nearly identical aircraft.
General 3 Wing area in square feet.
General 4 Wing loading in pounds per square foot based on normal take-off weight (but see General 2 for fighter-bombers).
General 5 Service ceiling in feet. (This heading omitted for fighter-bombers and tank-busters as irrelevant.)
General 6 Rate of climb in feet per minute assessed as in Chapter 4.
Speed*1 Stall.
Speed 2 Combat cruise for all aircraft except level bombers, for which economical cruise is shown.
Speed 3 Maximum speed in level flight.
Speed 4 Terminal velocity. For fighter-bombers this is the same as in their fighter configuration.
Acceleration 1 Throttle set at open, speed at beginning of move below maximum speed in level flight, in miles per hour per move.
Acceleration 2 Speed gain due to diving, given in miles per hour per centimetre (or inch) of height lost. (NB Each centimetre [or inch] represents 120 feet.)
Deceleration 1 Speed loss due to climbing, given in miles per hour per centimetre (or inch) of height gained.
Deceleration 2 Throttle set at open, speed at beginning of move above maximum speed in level flight, in miles per hour per move.
Deceleration 3 Throttle set at cruising, speed at beginning of move above maximum speed in level flight, in miles per hour per move. Also throttle set at closed, speed at beginning of move below maximum speed in level flight, in miles per hour per move.
Deceleration 4 Throttle set at closed, speed at beginning of move above maximum speed in level flight, in miles per hour per move.
Deceleration 5 Recoil due to firing fixed armament (not rockets) in miles per hour for each second of firing.
Turning ability Directly related to wing loading as Fig 4, given as turn radius/maximum speed at which this applies. For example, an aircraft with a wing loading of 43 lb per square foot would be shown as: 4/154
5/290
6/406
7/502.

*Speeds all in miles per hour.

The table-top game

This aircraft would turn on radius 4 at all speeds up to 154 mph, then radius 5 at speeds from 155 up to 290 mph, then on radius 6 etc. This does not apply to level bombers which turn on the radius shown at all speeds.
Fixed armament* 1 Cannon.
Fixed armament 2 Heavy machine-guns.
Fixed armament 3 Light machine-guns.
Traversing armament Due to the complexity of this it has been coded as follows: **Tu** = Turret; **S** = Swivel-mounted; **N** = Nose; **T** = Tail; **D** = Dorsal; **M** = Mid-upper; **V** = Ventral; **B** = Beam; **F** = Forward-firing; **R** = Rearward-firing; **C** = Cannon; **H** = Heavy machine-gun; and **L** = Light machine-gun.

For example, MTu2L would be a mid-upper turret with two light machine-guns worth one point each, while SVR2H would be a heavy machine-gun, swivel-mounted in a ventral position facing rearwards, and worth two points per second of firing.

In our two-dimensional vertical game beam guns cannot be used. Guns proved in action to be either totally useless or in any other way a liability have been omitted. These include rearward firing fixed armament and chin turrets, also the ventral dustbin type turret which, when extended, caused more drag than its firepower was worth. The arcs of fire from the various positions have been arbitrarily assessed as follows:
Nose and tail turrets 45 degrees up, 60 degrees down.
Nose and tail swivels 30 degrees up and 30 degrees down.
Mid-upper turrets—level fire to front through a complete overhead arc to level fire to rear.
Ventral turrets as mid-upper but underneath arc.
Dorsal turrets and swivels—level fire to 30 degrees up.
Ventral swivels—level fire to 30 degrees down.
Gun positions classified **F** or **R** cover the above arcs of fire but facing forward or rear respectively.
Defence values To save space, these are classified by letters A, B, C, etc. The details of these classifications are given hereunder.

	A	B	C	D	E	F	G	H	I	J	K	L	M	N	O	P	Q	R	S	T	U
Engines	7	8	8	7	7	8	9	7	8	7	8	9	9	8	8	8	7	8	6	8	8
Pilot	4	4	4	4	6	6	8	6	6	6	6	8	8	4	6	6	6	6	6	4	6
Oil	1	1	1	1	1	1	1	1	1	1	1	1	1	1	1	1	1	1	1	1	1
Fuel	2	2	2	3	3	3	4	3	3	5	5	4	4	2	5	7	9	9	9	4	5
Hydraulics	–	–	4	4	5	5	5	5	5	7	7	5	5	4	7	7	9	9	9	7	7
Wings	8	8	8	10	10	10	10	10	10	20	20	10	10	10	22	25	30	30	30	16	16
Fuselage	8	8	8	10	10	10	10	10	10	20	20	12	12	10	22	25	30	30	30	16	16
Tail	6	6	6	8	8	8	9	8	10	10	8	8	8	10	12	15	15	15	8	8	

Bomb load Is the weight in pounds of the normal load carried except fighter-bombers in which the load given is the maximum.
Rockets The number of rockets carried is given irrespective of weight or type.
Crew The normal crew. In practice this could vary quite a lot.
Torpedoes The number of torpedoes carried irrespective of weight or type. Where this is given for a bomber aircraft it means that one of the variants carried a torpedo load as an alternative to bombs. Other alternative loads could be mines or depth charges; these are not listed.

*Fixed and traversing armament given in points value per one second of firing.

Tank-busting aircraft have a heading of their own for large gun armament. No points value is given for these as they need to be tied in with whatever ground warfare rules you are using, although suggestions for this are made elsewhere.

The heading of fighters is self-explanatory, covering aircraft engaged in the air superiority role. Data for 50 fighter types is given, covering seven nationalities. The selection was governed mainly by the quantities manufactured but even with this number the problem was what to leave out.

The selection of fighter-bombers was, in a way, even more difficult as nearly all the aircraft in the fighter section carried bombs at one time or another. 14 were therefore selected as being reasonably representative.

Gun-armed tank-busters were relatively easy as there were not many to choose from. This category was extremely specialised and most nations preferred the versatility of the fighter-bomber. In consequence, the tank-buster was only used in those theatres of war where wide open spaces made conditions ideal for fast moving armoured warfare, namely Russia and North Africa. The Luftwaffe used the Henschel Hs 129B and the Junkers Ju 87G in Russia; the V-VS used the Il-2. Not many tank-busters were used in North Africa; a few Henschels were in service but were only in action over Tunisia, as in the Western Desert their appalling serviceability rate precluded their operational use. The RAF used the Hurricane IID which was the standard version but armed with two 40 mm cannon.

The type of armament and armour carried is of some interest as, in these low-level operations, the aircraft attracted a lot of ground fire, and in two cases were exceptionally well protected, while their guns had to be capable of puncturing the armour of a tank. To assist in this it was standard procedure to attack a tank from the rear where the amour was thinnest.

The Hurricane IID had extra armour protection for the engine and pilot, and retained two .303 Browning machine-guns in the wings, although these were used to assist in aiming the two 40 mm cannon rather than as offensive weaponry. The cannon, with a very slow rate of fire, were either Rolls-Royce BF with 12 rounds per gun or Vickers Type S with 15 rounds per gun, mounted in detachable gondolas under the wings. The Hurricane IV which was used in Burma also had provision for the 40 mm cannon as an alternative load, but as there were few Japanese tanks around at the time and the terrain favoured concealment, bombs and rockets were usually preferred.

The Henschel Hs 129B was the only German aircraft designed for the anti-tank role. A twin-engined, ugly, underpowered aircraft, it was massively armoured in the cockpit area and, as the war progressed, it carried larger and larger guns. The B1/R1 variant carried two 20 mm cannon, which were inadequate against anything but the lightest armour. This was followed in mid-1942 by the B1/R2 with a 30 mm MK 101 cannon with 30 rounds. This was an improvement, but it was incapable of penetrating the frontal armour of the Russian KV-1 and T-34 tanks. The B2/R3 variant carried a 37 mm weapon, which was still virtually useless in a frontal attack, although effective from other directions. This was followed in 1944 by the Hs 129B-3 armed with a 75 mm BK 75 cannon, with 12 shells which could be fired at the rate of two every three seconds. This was effective against almost all tank armour but the recoil was so vicious that a sustained firing pass was out of the question. One Hs 129 was even experimentally equipped with a flamethrower. The mind boggles.

The other tank-busting aircraft used by the Luftwaffe was the Junkers Ju 87 dive-bomber in its G variant with two 37 mm cannon slung under the wings in gondolas.

The table-top game 95

While very effective in the hands of an experienced pilot, handling in the air was adversely affected with the result that virtually no extra protection could be fitted as the weight penalty would have been unacceptable.

The Russian 'can-opener' was the most famous of all, the Il-2 Shturmovik. It was massively armoured, with protection for the engine and fuel tanks as well as the crew, the total weight of protection being not much less than one tonne. Two main versions were produced: the Il-2 single-seater early in the war, and the two-seater Il-2m3 which featured a rear gunner armed with a single 12.7 mm UBT machine-gun. The single seater carried two 20 mm ShVak cannon, which were fairly effective against the lighter German armour in 1941; and two 7.62 mm ShKAS machine-guns. It could also carry eight RS 82 rockets or an 880 lb bomb load. The two seater Il-2m3 entered service in the autumn of 1942, and had two 23 mm VYa cannon in place of the two ShVaks, which was a considerable improvement. In the spring of 1943 it was again up-gunned, carrying two 37 mm N37 cannon, which were capable of penetrating even the heavily armoured Panther and Tiger tanks. Finally, in this section only, maximum loaded weights have been used in the calculations.

The specialist dive-bomber carried dive brakes so that, alone among aircraft, it did not necessarily gain speed in a steep dive. This controlled diving speed naturally varied between aircraft types but to save unnecessary complications, if we say all dive-bombers using their dive brakes dive at a speed of 280 mph we will not be far wrong. All previous comments apply but, while after the bombs are dropped the aircraft should become more manoeuvreable due to the decreased wing loading, I prefer to ignore this. It will be partially compensated by not reducing the dive acceleration factor anyway. Certain dive-bombers were also used for level bombing, but the fact that they were stressed for high-g manoeuvres necessitates their inclusion in this section rather than elsewhere.

The single-engined torpedo bomber category was singled out for separate treatment for much the same reason as the dive-bombing category. They were generally agile aircraft and to categorise them with, say, the Heinkel 115 floatplane, would be rather unfair. Other loads, either bombs or mines, could be carried.

Our final category is the level bomber, the turning ability of which is standardised according to the number of engines (and thus, indirectly, to size). A couple of odd birds in the way of flying boats have crept in here. In self defence, it seemed a good place to put them at the time. Performance calculations are based on the normal loaded weight in all cases rather than the maximum permitted over-load.

Performance data—fighters

		Spitfire I	British Hurricane I	Gladiator II	Spitfire VB
General	1	1939	1939	1939	1941
	2	5,784	6,600	4,790	6,785
	3	242	258	323	242
	4	24	26	15	28
	5	34,000	33,200	33,000	37,000
	6	2,419	2,381	2,308	2,667

Air Battles in Miniature

British continued

		Spitfire I	Hurricane I	Gladiator II	Spitfire VB
Speeds	1	65	67	56	69
	2	305	266	205	324
	3	355	316	246	374
	4	444	395	308	468
Acceleration	1	+36	+32	+25	+37
	2	+ 3	+ 3	+ 2	+ 3
Deceleration	1	−15	−13	−11	−14
	2	− 9	− 8	− 6	− 9
	3	−18	−16	−13	−19
	4	−36	−32	−25	−37
	5	− 2	− 2	− 1	− 4
Turning ability		2/139	3/251	2/274	3/225
		3/277	4/373	3/308	4/351
		4/395	5/395	—	5/457
		5/444	—	—	6/468
Fixed armament	1	—	—	—	16
	2	—	—	—	—
	3	8	8	4	4
Trav. guns		—	—	—	—
Defence value		E	E	B	E
Crew		1	1	1	1

British continued

		Hurricane IIC	Spitfire IX	Typhoon IB	Beaufighter Mk VI F
General	1	1941	1942	1942	1942
	2	7,800	7,500	11,250	21,600
	3	258	242	279	503
	4	30	31	40	43
	5	35,600	44,000	35,200	26,500
	6	2,666	2,985	2,542	1,923
Speeds	1	71	72	80	83
	2	289	358	362	283
	3	339	408	412	333
	4	424	510	515	416
Acceleration	1	+34	+41	+41	+33
	2	+ 3	+ 3	+ 5	+ 5
Deceleration	1	−13	−14	−16	−17
	2	− 9	−11	−10	− 8
	3	−17	−21	−20	−17
	4	−34	−41	−41	−33
	5	− 7	− 4	− 4	− 3
Turning ability		3/199	3/184	4/199	4/154
		4/329	4/316	5/329	5/290
		5/424	5/428	6/439	6/406
		—	6/510	7/515	7/416

The table-top game

		Hurricane IIC	Spitfire IX	Typhoon IB	Beaufighter Mk VI F
		British continued			
Fixed armament	1	32	16	32	32
	2	—	—	—	—
	3	—	4	—	6
Trav. guns		—	—	—	—
Defence value		E	E	E	K
Crew		1	1	1	2

		Mosquito Mk VI	Spitfire XIV	Tempest V	Morane MS 406
		British continued			*French*
General	1	1943	1944	1944	1939
	2	22,300	8,500	11,500	5,610
	3	435	242	302	172
	4	51	35	38	33
	5	36,000	44,500	36,000	32,808
	6	2,142	2,857	3,000	2,175
Speeds	1	91	76	78	74
	2	330	398	377	254
	3	380	448	427	304
	4	475	560	534	380
Acceleration	1	+38	+45	+43	+30
	2	+ 5	+ 4	+ 5	+ 3
Deceleration	1	−18	−16	−14	−14
	2	−10	−11	−11	− 8
	3	−19	−23	−22	−15
	4	−38	−45	−43	−30
	5	− 2	− 4	− 4	− 3
Turning ability		5/184	4/264	4/225	3/154
		6/316	5/384	5/351	4/290
		7/428	6/484	6/457	5/380
		8/475	7/560	7/534	—
Fixed armament	1	32	16	32	8
	2	—	—	—	—
	3	4	4	—	2
Trav. guns		—	—	—	—
Defence value		J	E	E	D
Crew		2	1	1	1

		Dewoitine D 520	Messerschmitt Bf 109E-3	Messerschmitt Bf 110C-4	Messerschmitt Bf 109F-3
		French continued	*German*		
General	1	1940	1939	1939	1941
	2	5,900	5,523	14,884	6,063

		French continued Dewoitine D 520	German continued Messerschmitt Bf 109E-3	Messerschmitt Bf 110C-4	Messerschmitt Bf 109F-3
General	3	172	174	413	173
	4	34	32	36	35
	5	33,628	36,091	32,811	39,370
	6	2,187	2,646	1,200	3,462
Speeds	1	75	73	76	76
	2	282	304	299	341
	3	332	354	349	391
	4	415	443	436	489
Acceleration	1	+33	+35	+35	+39
	2	+ 3	+ 2	+ 3	+ 3
Deceleration	1	−15	−13	−22	−11
	2	− 8	− 9	− 9	−10
	3	−16	−18	−18	−20
	4	−33	−35	−35	−39
	5	− 3	− 2	− 1	− 1
Turning ability		4/277	3/169	4/251	4/264
		5/395	4/303	5/373	5/384
		6/415	5/417	6/436	6/484
		—	6/443	—	7/489
Fixed armament	1	8	6	6	4
	2	—	—	—	—
	3	4	2	4	2
Trav. guns		—	—	DSR 1 L	—
Defence value		E	E	J	E
Crew		1	1	2	1

		Junkers Ju 88C-4	German continued Focke-Wulf FW 190A-3	Messerschmitt Bf 109G-6	Focke-Wulf FW 190A-8
General	1	1941	1942	1943	1944
	2	26,600	8,000	6,945	9,750
	3	587	197	173	197
	4	45	41	40	49
	5	26,900	37,403	38,551	37,403
	6	1,800	3,474	3,311	2,020
Speeds	1	85	81	80	89
	2	248	344	337	358
	3	298	394	387	408
	4	373	493	484	510
Acceleration	1	+30	+39	+39	+41
	2	+ 6	+ 4	+ 3	+ 4
Deceleration	1	−17	−11	−12	−20
	2	− 8	−10	−10	−10
	3	−15	−20	−20	−21

The table-top game 99

German continued

		Junkers Ju 88C-4	Focke-Wulf FW 190A-3	Messerschmitt Bf 109G-6	Focke-Wulf FW 190A-8
Deceleration	4	−30	−39	−39	−41
	5	− 1	− 4	− 6	− 4
Turning ability		5/264	4/184	4/199	5/212
		6/373	5/316	5/329	6/340
		—	6/428	6/439	7/448
		—	7/493	7/484	8/510
Fixed armament	1	6	20	34	28
	2	—	—	4	4
	3	4	2	—	—
Trav. guns		DSR 1 L	—	—	—
Defence value		J	F	E	G
Crew		3	1	1	1

German continued | Italian

		Focke-Wulf FW 190D-9	Messerschmitt Me 262A-1a	Fiat CR 32 Quater	Fiat CR 42 Falco
General	1	1944	1944	1940	1940
	2	9,480	14,101	4,222	5,033
	3	197	234	238	252
	4	48	60	18	20
	5	39,372	36,080	25,262	33,465
	6	2,773	3,937	1,360	2,364
Speeds	1	88	100	59	61
	2	376	486	183	214
	3	426	536	220	257
	4	533	590	275	321
Acceleration	1	+43	+54	+22	+26
	2	+ 4	+ 3	+ 2	+ 2
Deceleration	1	−16	−14	−16	−11
	2	−11	−14	− 6	− 7
	3	−22	−27	−11	−13
	4	−43	−54	−22	−26
	5	− 3	− 5	− 1	− 1
Turning ability		5/225	6/199	2/229	2/199
		6/351	7/329	3/275	3/321
		7/457	8/439	—	—
		8/533	9/529	—	—
		—	10/590	—	—
Fixed armament	1	14	80	—	—
	2	4	—	4	4
	3	—	—	—	—
Trav. guns		—	—	—	—
Defence value		E	I	A	B
Crew		1	1	1	1

The natural unit of fighter aircraft is the pair; the leader and his No 2 who guards his tail. Sometimes a single aircraft was used as live bait, with covering aircraft at high level. Watch this as an example of how not to use both systems. **1** Two Bf 109s out for a stroll, espy a fat, juicy Hurricane stooging about below. He also sees them. **2** Nose down and away we go lads. Perfectly set up for the British high cover? Would be, except the Hurricane *is* the British high cover. **3** The dive continues and the '109s, with their superior acceleration, are overhauling the Hurricane. The '109 leader fires and knocks lumps out of the Hurricane's fuselage. The No 2 is also closing up for a shot, and no longer covering his leader's tail.

The table-top game   101

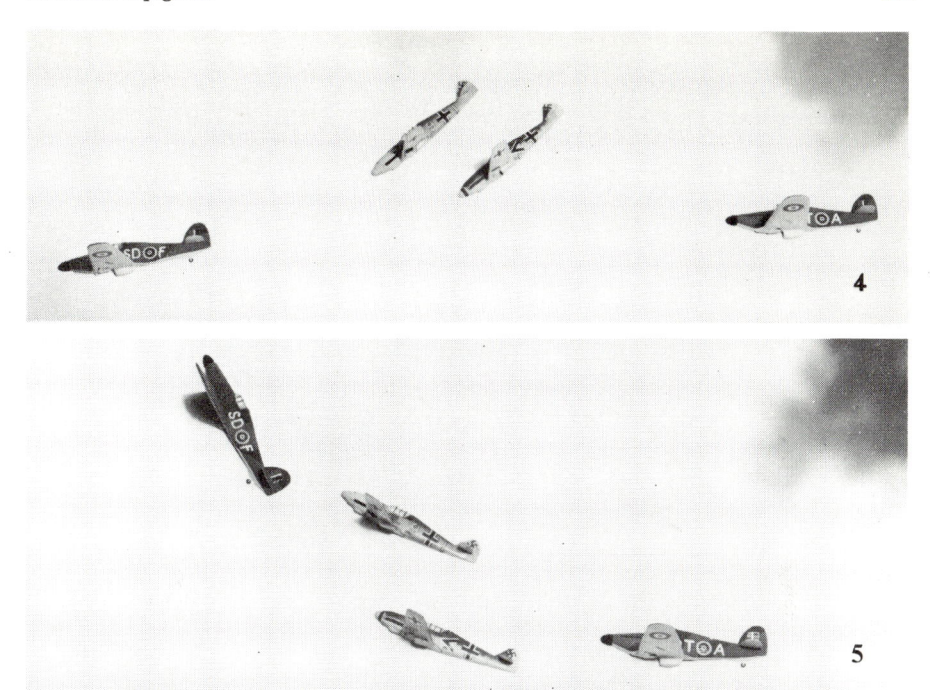

4 Here comes the cavalry. The second Hurricane sneaks in unobserved as the first one levels out. The '109s are really closing fast now. **5** The lead Hurricane breaks upwards, but to no avail; the '109 leader cuts across the turn and riddles his tail. Meanwhile the second '109 has paid the penalty of being out of position, getting a cockpit full of lead from the second Hurricane. **6** The first Hurricane and the second Messerschmitt both go down out of control; the '109 leader, suddenly aware of the second Hurricane, breaks hard upwards. **7** Converting the speed from his dive into height, the '109 leader climbs away, easily outdistancing the second Hurricane.

		Fiat G50bis Freccia	Italian continued Macchi C200 Saetta	Macchi C202 Folgore	Russian Polikarpov I-16 Type 24
General	1	1940	1940	1942	1941
	2	5,512	5,132	6,459	4,215
	3	224	181	181	161
	4	25	28	36	26
	5	35,270	29,200	37,730	29,530
	6	2,402	2,795	3,934	2,828
Speeds	1	66	69	76	67
	2	252	263	320	276
	3	302	313	370	326
	4	378	391	463	408
Acceleration	1	+30	+31	+37	+33
	2	+ 2	+ 2	+ 3	+ 2
Deceleration	1	−12	−11	− 9	−12
	2	− 8	− 8	− 9	− 8
	3	−15	−16	−19	−17
	4	−30	−31	−37	−33
	5	− 1	− 1	− 1	− 6
Turning ability		3/264	3/225	4/251	3/251
		4/378	4/351	5/373	4/373
		—	5/391	6/463	5/408
		—	—	—	—
Fixed armament	1	—	—	—	14
	2	4	4	4	—
	3	—	—	2	2
Trav. guns		—	—	—	—
Defence value		F	F	E	F
Crew		1	1	1	1

		LaGG 3	Russian continued MiG 3	Yak 1M	La 5FN
General	1	1941	1941	1942	1943
	2	7,032	7,242	5,730	7,407
	3	188	189	160	188
	4	37	38	36	39
	5	31,495	39,370	35,335	31,168
	6	2,803	2,877	3,645	3,489
Speeds	1	77	78	76	79
	2	298	347	330	353
	3	348	397	380	403
	4	435	496	475	504
Acceleration	1	+35	+40	+38	+40
	2	+ 3	+ 3	+ 3	+ 3
Deceleration	1	−12	−14	−10	−11
	2	− 9	−10	−10	−10

The table-top game

		Russian continued			
		LaGG3	MiG 3	Yak 1M	La 5FN
Deceleration	3	—18	—20	—19	—20
	4	—35	—40	—38	—40
	5	— 2	— 1	— 2	— 3
Turning ability		4/238	4/225	4/251	4/212
		5/362	5/351	5/373	5/340
		6/435	6/457	6/475	6/448
		—	7/496	—	7/504
Fixed armament	1	7	—	7	14
	2	4	2	2	—
	3	—	2	—	—
Trav. guns		—	—	—	—
Defence value		E	E	E	F
Crew		1	1	1	1

		Russian continued		American	
		Yak 9D	Yak 3	P-39D Airacobra	P-40C Warhawk
General	1	1943	1944	1942	1942
	2	6,867	5,864	7,650	7,459
	3	185	160	213	236
	4	37	37	36	32
	5	32,810	35,430	32,100	29,500
	6	3,348	4,000	2,160	1,667
Speeds	1	77	77	76	73
	2	323	357	318	295
	3	373	407	368	345
	4	466	509	448	431
Acceleration	1	+37	+41	+37	+35
	2	+ 3	+ 3	+ 3	+ 3
Deceleration	1	—11	—10	—18	—21
	2	— 9	—10	—10	— 9
	3	—19	—21	—19	—18
	4	—37	—41	—37	—35
	5	— 2	— 1	— 4	— 1
Turning ability		4/238	4/238	4/251	3/169
		5/362	5/362	5/373	4/303
		6/466	6/466	6/448	5/417
		—	7/509	—	6/431
Fixed armament	1	7	—	20	—
	2	2	4	4	4
	3	—	—	4	4
Trav. guns		—	—	—	—
Defence value		E	E	E	E
Crew		1	1	1	1

Air Battles in Miniature

		F4F3 Wildcat	American continued P-38J Lightning	P-51B Mustang	F6F3 Hellcat
General	1	1942	1943	1943	1943
	2	7,002	17,500	10,100	12,186
	3	260	328	233	334
	4	27	53	43	36
	5	37,500	44,000	42,500	37,500
	6	1,786	2,850	2,600	1,948
Speeds	1	68	93	83	76
	2	280	364	387	326
	3	330	414	437	376
	4	413	518	546	470
Acceleration	1	+33	+41	+44	+38
	2	+ 3	+ 4	+ 5	+ 5
Deceleration	1	−18	−14	−17	−20
	2	− 8	−10	−11	−10
	3	−17	−21	−22	−19
	4	−33	−41	−44	−38
	5	− 2	− 1	− 2	− 1
Turning ability		3/238	6/290	4/154	4/251
		4/362	7/406	5/290	5/373
		5/413	8/502	6/406	6/470
		—	9/518	7/502	—
		—	—	8/546	—
Fixed armament	1	—	7	—	—
	2	8	8	12	12
	3	—	—	—	—
Trav. guns		—	—	—	—
Defence value		F	H	E	F
Crew		1	1	1	1

		American continued F4U1A Corsair	P-47D Thunderbolt	Japanese Nakajima Ki 43-Ic Hayabusa	Mitsubishi A6M2 Type 0 Model 21 Zero-Sen
General	1	1943	1944	1942	1942
	2	11,149	14,600	4,360	5,313
	3	314	308	237	242
	4	36	47	18	22
	5	37,900	40,000	33,420	32,810
	6	2,381	1,739	3,395	2,630
Speeds	1	76	87	59	63
	2	345	377	254	267
	3	395	427	304	317
	4	494	534	380	396

The table-top game

		American continued		Japanese continued	
		F4U1A Corsair	P-47D Thunderbolt	Nakajima Ki 43-Ic Hayabusa	Mitsubishi A6M2 Type 0 Model 21 Zero-Sen
Acceleration	1	+40	+43	+30	+32
	2	+ 5	+ 7	+ 2	+ 2
Deceleration	1	−17	−24	− 9	−12
	2	−10	−11	− 8	− 8
	3	−20	−22	−15	−16
	4	−40	−43	−30	−32
	5	− 2	− 2	− 1	− 2
Turning ability		4/251	5/238	2/229	2/169
		5/373	6/362	3/351	3/303
		6/475	7/466	4/380	4/396
		7/494	8/534	—	—
Fixed armament	1	—	—	—	6
	2	12	16	—	—
	3	—	—	2	2
Trav. guns		—	—	—	—
Defence value		F	F	C	C
Crew		1	1	1	1

		Japanese continued			
		Nakajima Ki 44-1 Type 2 Shoki	Nakajima Ki 43-I-2B Hayabusa	Kawasaki Ki 61-Ia Type 3 Hien	Mitsubishi A6M6C Type 0 Model 53c Zero-Sen
General	1	1942	1943	1943	1944
	2	5,512	5,320	7,650	6,026
	3	180	237	215	229
	4	31	22	36	26
	5	35,500	36,800	32,800	35,100
	6	2,500	2,819	2,343	2,564
Speeds	1	72	63	76	67
	2	310	270	298	296
	3	360	320	348	346
	4	450	400	435	433
Acceleration	1	+36	+32	+35	+35
	2	+ 2	+ 2	+ 3	+ 3
Deceleration	1	−14	−11	−15	−14
	2	− 9	− 8	− 9	− 9
	3	−18	−16	−18	−18
	4	−36	−32	−35	−35
	5	− 2	− 1	− 3	− 3
Turning ability		3/184	2/169	4/251	3/251
		4/316	3/303	5/373	4/373

		Nakajima Ki 44-1 Type 2 Shoki	Japanese continued Nakajima Ki 43-I-2B Hayabusa	Kawasaki Ki 61-Ia Type 3 Hien	Mitsubishi A6M6C Type 0 Model 53c Zero-Sen
Turning ability		5/428 6/450	4/400 —	6/435 —	5/433 —
Fixed armament	1	—	—	14	6
	2	4	4	—	2
	3	2	—	2	1
Trav. guns		—	—	—	—
Defence value		C	F	E	F
Crew		1	1	1	1

		Japanese continued Nakajima Ki 84-1a Type 4 Model 1A Hayate	Kawanishi N1K2J Shiden 21
General	1	1944	1944
	2	7,965	9,039
	3	226	253
	4	35	36
	5	38,000	35,300
	6	2,780	2,672
Speeds	1	76	76
	2	338	320
	3	388	370
	4	484	463
Acceleration	1	+39	+37
	2	+ 4	+ 4
Deceleration	1	−14	−14
	2	−10	− 9
	3	−20	−19
	4	−39	−37
	5	− 3	− 2
Turning ability		4/264 5/384 6/484 —	4/251 5/373 6/463 —
Fixed armament	1	14	12
	2	4	—
	3	—	—
Trav. guns		—	—
Defence value		F	F
Crew		1	1

The table-top game

Performance data—fighter bombers

		Hurricane IIC	British Spitfire IX	Typhoon IB	German Messerschmitt Bf 109G-6
General	1	1941	1942	1942	1943
	2	8,800	8,500	13,250	7,495
	3	258	242	279	173
	4	34	35	47	43
	6	2,133	2,388	2,034	2,649
Speeds	1	75	76	87	83
	2	241	308	312	287
	3	289	358	362	337
	4	424	510	515	484
Acceleration	1	+29	+36	+36	+34
	2	+ 4	+ 4	+ 6	+ 3
Deceleration	1	−14	−15	−18	−13
	2	− 7	− 8	− 9	− 9
	3	−14	−18	−18	−17
	4	−29	−36	−36	−34
	5	− 7	− 4	− 4	− 6
Turning ability		4/277	4/264	5/238	4/154
		5/395	5/384	6/362	5/290
		6/424	6/484	7/466	6/406
		—	7/510	8/515	7/484
Fixed armament	1	32	16	32	34
	2	—	—	—	4
	3	—	4	—	—
Trav. guns		—	—	—	—
Defence value		E	E	E	E
Crew		1	1	1	1
Bomb load		1,000	1,000	2,000	550
Rockets		8	—	8	2 (21 cm)

		German continued Focke-Wulf FW 190A-8	Fiat CR 42 Falco	Italian Macchi C200 Saetta	Russian LaGG3
General	1	1944	1940	1940	1941
	2	11,950	5,473	5,838	7,472
	3	197	252	181	188
	4	61	22	32	40
	6	1,616	1,827	2,236	2,242
Speeds	1	101	63	73	80
	2	308	164	213	248
	3	358	214	263	298
	4	510	321	391	435
Acceleration	1	+36	+21	+26	+30
	2	+ 5	+ 2	+ 3	+ 3

		German continued Focke-Wulf FW 190A-8	**Italian continued** Fiat CR 42 Falco	Macchi C200 Saetta	**Russian continued** LaGG3
Deceleration	1	—22	—11	—12	—13
	2	— 9	— 5	—7	— 7
	3	—18	—11	—13	—15
	4	—36	—21	—26	—30
	5	— 4	— 1	— 1	— 2
Turning ability		6/184	2/169	3/169	4/199
		7/316	3/303	4/303	5/329
		8/428	4/321	5/391	6/435
		9/510	—	—	—
Fixed armament	1	28	—	—	7
	2	4	4	4	4
	3	—	—	—	—
Trav. guns		—	—	—	—
Defence value		G	B	F	E
Crew		1	1	1	1
Bomb load		2,200	440	706	440
Rockets		—	—	—	4

		Russian continued La 5 FN	P-40C Warhawk	**American** P-38J Lightning	P-47D Thunderbolt
General	1	1943	1942	1943	1944
	2	8,067	7,959	19,500	16,600
	3	188	236	328	308
	4	43	34	59	54
	6	2,791	1,334	2,280	1,391
Speeds	1	83	75	99	94
	2	303	245	314	327
	3	353	295	364	377
	4	504	431	518	534
Acceleration	1	+35	+30	+36	+38
	2	+ 4	+ 4	+ 4	+ 7
Deceleration	1	—13	—22	—16	—27
	2	— 9	— 7	— 9	— 9
	3	—18	—15	—18	—19
	4	—35	—30	—36	—38
	5	— 3	— 1	— 1	— 2
Turning ability		4/154	4/277	6/212	6/277
		5/290	5/395	7/340	7/395
		6/406	6/431	8/448	8/493
		7/502	—	9/518	9/534
		8/504	—	—	—

The table-top game

		Russian continued La 5FN	P-40C Warhawk	American continued P-38J Lightning	P-47D Thunderbolt
Fixed armament	1	14	—	7	—
	2	—	4	8	16
	3	—	4	—	—
Trav. guns		—	—	—	—
Defence value		F	E	H	F
Crew		1	1	1	1
Bomb load		660	500	2,000	2,000
Rockets		4	—	10	6

		Ki 43-I-2B Hayabusa	Japanese A6M6C Type 0 Model 53C Zero-Sen
General	1	1942	1944
	2	4,910	6,576
	3	237	229
	4	21	29
	6	2,716	2,051
Speeds	1	62	70
	2	204	256
	3	254	296
	4	380	433
Acceleration	1	+25	+30
	2	+ 2	+ 3
Deceleration	1	− 9	−15
	2	− 6	− 7
	3	−13	−15
	4	−25	−30
	5	− 1	− 3
Turning ability		2/184	3/212
		3/316	4/340
		4/380	5/433
		—	—
Fixed armament	1	—	6
	2	—	2
	3	2	1
Trav. guns		—	—
Defence value		C	F
Crew		1	1
Bomb load		550	550
Rockets		—	—

Performance data—tank busters

		British Hurricane IID	German Ju 87G-1	German Hs 129B2	Russian Il-2m3
General	1	1942	1943	1943	1942
	2	8,100	14,400	11,266	14,021
	3	258	343	312	414
	4	31	42	36	34
	6	1,613	1,232	1,406	1,547
Speeds	1	72	82	76	75
	2	238	207	211	209
	3	286	248	253	251
	4	358	310	316	314
Acceleration	1	+29	+25	+25	+25
	2	+ 4	+ 6	+ 5	+ 6
Deceleration	1	−18	−20	−18	−16
	2	− 7	− 6	− 6	− 6
	3	−14	−12	−13	−13
	4	−29	−25	−25	−25
	5	− 7	− 6	− 2	− 4
Turning ability		3/184	4/169	4/251	4/277
		4/316	5/303	5/316	5/314
		5/358	6/310	—	—
		—	—	—	—
Fixed armament					
Anti-tank		2 × 40 mm	2 × 37 mm	1 × 37 mm	2 × 23 mm
	2	—	—	4	—
	3	2	2	—	2
Trav. guns		—	DSR 2 L	—	DSR 2 H
Defence value		E	E	M	L
Crew		1	2	1	2
Bomb load		—	—	—	880
Rockets		—	—	—	8

Performance data—dive-bombers

		British Blackburn Skua	German Junkers Ju 87B-2	German Junkers Ju 88A-4	Russian Petlyakov Pe-2
General	1	1939	1939	1941	1941
	2	8,240	9,321	30,870	18,783
	3	248	343	590	436
	4	33	27	52	43
	5	15,200	26,248	27,880	29,530
	6	740	828	761	2,343
Speeds	1	74	68	92	83
	2	188	198	228	286

The table-top game 111

		British continued Blackburn Skua	German continued Junkers Ju 87B-2	Junkers Ju 88A-4	Russian continued Petlyakov Pe-2
Speeds	3	225	237	273	336
	4	281	296	341	420
Acceleration	1	+23	+24	+27	+34
	2	+ 4	+ 4	+ 7	+ 4
Deceleration	1	−31	−29	−35	−15
	2	− 6	− 6	− 7	− 8
	3	−11	−12	−14	−17
	4	−23	−24	−27	−34
	5	− 2	—	—	− 1
Turning ability		3/154	3/238	5/169	4/154
		4/281	4/296	6/303	5/290
		—	—	7/307	6/406
		—	—	—	7/420
Fixed armament	1	—	—	—	—
	2	—	—	2	2
	3	1	1	—	1
Trav. guns		DSR 1 L	DSR 2 L	DSR 1 L	DSR 1 L
				VSR 1 L	VSR 1 L
				BS 2 L	
Defence value		N	E	J	J
Crew		2	2	4	3
Bomb load		500	2,205	4,400	2,205

		American SBD 3 Dauntless	SB 2C1 Helldiver	Japanese Aichi D3 A1	Yokosuka D4Y Suisei
General	1	1942	1944	1942	1943
	2	9,519	15,076	8,047	8,610
	3	325	422	376	254
	4	29	36	21	34
	5	25,200	24,700	30,511	32,400
	6	1,429	1,236	1,526	2,174
Speeds	1	70	76	62	75
	2	213	234	200	311
	3	255	281	240	361
	4	319	351	300	451
Acceleration	1	+26	+28	+24	+36
	2	+ 4	+ 7	+ 4	+ 4
Deceleration	1	−18	−23	−16	−17
	2	− 6	− 7	− 6	− 9
	3	−13	−14	−12	−18
	4	−26	−28	−24	−36
	5	− 1	− 1	− 1	− 1

	American continued		Japanese continued	
	SBD-3 Dauntless	**SB 2C1 Helldiver**	**Aichi D3 A1**	**Yokosuka D4Y Suisei**
Turning ability	3/212	4/251	2/184	4/277
	4/319	5/351	3/300	5/395
	—	—	—	6/451
	—	—	—	—
Fixed armament 1	—	—	—	—
2	4	8	—	—
3	—	—	2	2
Trav. guns	DSR 2 H	DSR 2 H	DSR 1 L	DSR 1 L
Defence value	F	F	C	E
Crew	2	2	2	2
Bomb load	1,000	1,000	551	1,100

Performance data—single-engined torpedo bombers

		British		American	Japanese	
		Fairey Swordfish	**Fairey Barracuda II**	**TBM-1C Avenger**	**Nakajima B5N2**	**Nakajima B6N1 Tenzan**
General	1	1939	1943	1942	1942	1944
	2	7,510	16,400	17,364	9,030	11,464
	3	607	425	490	406	400
	4	12	39	35	22	29
	5	16,500	18,400	21,400	27,099	26,660
	6	665	1,740	769	1,284	1,562
Speeds	1	52	79	76	63	70
	2	115	220	214	196	250
	3	138	264	257	235	300
	4	173	330	321	294	375
Acceleration	1	+14	+26	+26	+24	+30
	2	+ 3	+ 7	+ 8	+ 4	+ 5
Deceleration	1	−21	−15	−34	−19	−19
	2	− 3	− 7	− 6	− 6	− 8
	3	− 7	−13	−13	−12	−15
	4	−14	−26	−26	−24	−30
	5	—	—	− 1	—	—
Turning ability	1	2/173	4/212	4/264	2/169	3/212
	2	—	5/330	5/321	3/294	4/340
	3	—	—	—	—	5/375
	4	—	—	—	—	—
Fixed armament	1	—	—	—	—	—
	2	—	—	4	—	—
	3	1	—	—	—	1
Trav. guns		DSR 1 L	DSR 2 H	DTuR 2 H	DSR 1 L	DSR 1 L
				VSR 2 H		VSR 1 L

The table-top game

	British continued		American continued	Japanese continued	
	Fairey Swordfish	Fairey Barracuda II	TBM-1C Avenger	Nakajima B5N2	Nakajima B6N1 Tenzan
Defence value	B	E	F	C	F
Crew	3	3	3	3	2
Torpedo	1	1	1	1	1

Performance data—bombers

		British			
		Blenheim IV	Wellington Ic	Sunderland I	Stirling I
General	1	1939	1939	1939	1941
	2	14,400	28,500	44,600	59,400
	3	469	840	1,487	1,460
	4	31	34	30	41
	5	22,000	18,000	17,000	20,500
	6	1,246	542	620	357
Speeds	1	72	75	71	81
	2	200	180	178	200
	3	266	235	210	260
	4	333	294	263	325
Acceleration	1	+27	+24	+21	+26
	2	+ 2	+ 3	+ 1	+ 2
Deceleration	1	−22	−44	−34	−73
	2	− 7	− 6	− 5	− 7
	3	−13	−12	−11	−13
	4	−27	−24	−21	−26
Turning ability		10/333	10/294	12/263	12/325
Fixed armament		1 L	—	—	—
Trav. guns		MTu 2L	NTu 2L	NTu 2L	NTu 2L
			BS 2L	BS 2L	MTu 2L
			TTu 2L	TTu 4L	TTu 4L
Defence value		K	O	R	R
Crew		3	6	10	8
Bomb load		1,000	4,500	2,000	14,000
Torpedo alternatives		—	1	—	—

		British continued		German	
		Mosquito IV	Lancaster I	Dornier Do 17Z-2	Heinkel He 111H
General	1	1942	1942	1939	1939
	2	20,870	68,000	18,940	25,000
	3	435	1,297	592	943
	4	48	52	32	27
	5	40,000	24,500	26,904	25,500

Air Battles in Miniature

		British continued		**German continued**	
		Mosquito IV	Lancaster I	Dornier Do 17Z-2	Heinkel He 111H
General	6	1,935	481	640	781
Speeds	1	88	92	73	68
	2	300	210	186	224
	3	380	287	255	258
	4	475	359	319	323
Acceleration	1	+38	+29	+26	+26
	2	+ 2	+ 2	+ 2	+ 3
Deceleration	1	−20	−60	−41	−33
	2	−10	− 7	− 6	− 6
	3	−19	−14	−13	−13
	4	−38	−29	−26	−26
Turning ability		10/475	12/359	10/319	10/323
Fixed armament		—	—	—	—
Trav. guns		—	NTu 2 L	NSF 1 L	NSF 1 L
			MTu 2 L	BS 2 L	BS 2 L
			TTu 4 L	DSR 1 L	DSR 1 L
				VSR 1 L	VSR 1 L
Defence value		J	Q	K	J
Crew		2	7	5	5
Bomb load		2,000	14,000	2,205	4,410
Torpedo alternatives		—	—	—	2

		German continued			**Italian**
		Heinkel He 115B-1	Heinkel He 177A-5	Arado Ar 234B-2	Savoia-Marchetti SM 79-II
General	1	1939	1943	1944	1940
	2	20,065	59,966	18,541	24,912
	3	935	1,098	284	657
	4	21	55	65	38
	5	18,040	26,250	32,808	22,966
	6	684	513	1,538	1,131
Speeds	1	62	95	105	78
	2	183	210	420	230
	3	220	303	461	270
	4	275	379	576	338
Acceleration	1	+22	+30	+46	+27
	2	+ 2	+ 2	+ 2	+ 1
Deceleration	1	−32	−58	−30	−24
	2	− 6	− 8	−12	− 7
	3	−11	−15	−23	−14
	4	−22	−30	−46	−27
Turning ability		10/275	12/379	10/576	11/338

The table-top game 115

	German continued			**Italian continued**
	Heinkel He 115B-1	Heinkel He 177A-5	Arado Ar 234B-2	Savoia-Marchetti SM 79-II
Fixed armament	—	—	—	—
Trav. guns	NSF 1 L	NSF 1 L	—	BS 1 L
	DSR 1 L	VSF 7 C		DSR 2 H
		VSR 2 L		VSR 1 L
		D Barbette R		
		4 H		
		DTu 2 H		
		TS 7 C		
Defence value	K	S	K	P
Crew	3	5	1	4
Bomb load	3,300	13,200	4,410	2,204
Torpedo alternatives	1	2	—	2

		Italian continued Cant Z-1007 bis	**Russian** Ilyushin Il-4	Tupolev Tu 2S	**American** B-25J Mitchell
General	1	1940	1942	1944	1942
	2	28,260	18,475	25,044	33,500
	3	854	718	525	610
	4	33	26	48	55
	5	26,500	31,820	31,170	25,000
	6	937	1,832	1,726	1,110
Speeds	1	74	67	88	95
	2	235	190	286	200
	3	280	267	342	275
	4	350	334	428	344
Acceleration	1	+28	+27	+34	+28
	2	+ 1	+ 2	+ 3	+ 4
Deceleration	1	−30	−15	−20	−25
	2	− 7	− 7	− 9	− 7
	3	−14	−13	−17	−14
	4	−28	−27	−34	−28
Turning ability		11/350	10/334	10/428	10/344
Fixed armament		—	—	14 C	4 H
Trav. guns		BS 2 L	NSF 1 L	DSR 2 H	NS 2 H
		DTu 2 H	MTu 2 H	VSR 2 H	MTu 4 H
		VSR 2 H	VSR 1 L		BS 4 H
					TTu 4 H
Defence value		P	K	K	K
Crew		5	3	4	5
Bomb load		4,408	2,204	2,204	3,000
Torpedo alternatives		2	1	—	1

		B-26G Marauder	American continued B-24J Liberator	B-17G Fortress	PBY 5 Catalina
General	1	1942	1942	1942	1942
	2	37,000	56,000	55,000	26,200
	3	658	1,048	1,420	1,400
	4	56	53	39	19
	5	28,000	28,000	35,000	18,100
	6	1,000	800	541	800
Speeds	1	96	93	79	60
	2	232	215	160	115
	3	305	300	287	189
	4	381	375	359	236
Acceleration	1	+31	+30	+29	+19
	2	+ 4	+ 2	+ 2	+ 3
Deceleration	1	−31	−38	−54	−24
	2	− 8	− 8	− 7	− 5
	3	−15	−15	−14	− 9
	4	−31	−30	−29	−19
Turning ability		10/381	12/375	12/359	10/236
Fixed armament		2 H	—	—	—
Trav. guns		NS 2H	NTu 4H	NTu 4H	NS 1L
		MTu 4H	MTu 4H	MTu 4H	BS 4H
		BS 4H	VTu 4H	VTu 4H	VSR 1L
		TTu 4H	BS 4H	BS 8H	
			TTu 4H	TTu 4H	
Defence value		K	R	R	K
Crew		5	9	10	7
Bomb load		4,000	12,800	17,600	4,000
Torpedo alternatives		1	—	—	2

		Mitsubishi G4M2a Type 1 Model 24 [Betty]	Japanese Mitsubishi Ki 21 Type 97 Model 2B [Sally]	Mitsubishi Ki 67 Type 4 Model 1B [Peggy]
General	1	1942	1942	1944
	2	27,557	21,407	30,346
	3	841	749	741
	4	33	29	41
	5	29,360	32,800	31,070
	6	1,229	1,496	1,358
Speeds	1	74	70	81
	2	196	176	248
	3	272	297	334
	4	340	371	418

The table-top game

		Mitsubishi G4M2a Type 1 Model 24 [Betty]	Japanese continued Mitsubishi Ki 21 Type 97 Model 2B [Sally]	Mitsubishi Ki 67 Type 4 Model 1B [Peggy]
Acceleration	1	+27	+30	+33
	2	+ 3	+ 2	+ 3
Deceleration	1	−22	−20	−24
	2	− 7	− 7	− 8
	3	−14	−15	−17
	4	−27	−30	−33
Turning ability		10/340	10/297	10/334
Fixed armament		—	—	—
Trav. guns		NS 1 L	NS 1 L	NS 2 H
		DTu 3 C	DTu 2 H	DSR 7 C
		BS 6 C	VSR 1 L	BS 4 H
		TS 3 C	BS 2 L	TS 2 H
			TS 1 L	
Defence value		T	T	U
Crew		7	5	6
Bomb load		2,200	2,200	1,760
Torpedo alternatives		1	—	1

Rules for play

Wargamers are, in my experience, an argumentative lot. As we want a game and not an argument, rule one is most important.

1 The game is to be played at all times in a spirit of sporting competition. Any circumstance not covered somewhere in this volume is to be settled both reasonably and amicably. Failure to comply with this rule will bring down the curse of the Gremlins on your unhappy heads. May they load your guns with washing-up liquid and cocoa. It has happened! Extract from combat report dated February 29 1942: 'I gave him a three-second squirt and he went down covered in dark brown lather'.

2 All orders are to be clearly written before the move commences, and should cover movement, throttle setting and bomb-dropping.

3 All observation, movement and firing is simultaneous *except as defined hereafter*.

Game sequence

4 Measure and record height above the baseline of the playing surface in centimetres (or inches). Measurement to be baseline to nose of model.

5 Resolve observation as Chapter 5.

6 Write orders. This should cover movement, throttle setting and intended dropping of bombs. Firing is at targets of opportunity and written orders are not necessary. A time limit should be set of one minute maximum. Orders must be clear and exact. For ease I suggest standard abbreviations are used as follows: throttle settings—O = open, C = cruise and S = shut; movement—Str = continue present course, LT = commence level turn, U = upwards, D = downwards and R = radius.

A combination of these might be U60° R4 Str, which would mean that the aircraft turns upwards an angle of 60° on radius 4, then completes its remaining move distance in a straight line. Due to the short time scale S-turns cannot be made. If attacking a ground target, when only a small adjustment of course is necessary, ie, less than 15 degrees, it is sufficient to write 'attack', stating the target. If dropping bombs from less than a 60 degree dive, the exact point of bomb release must be written beforehand, for example BR at 122 mm.

No on table measurement prior to writing orders is permitted!

One other order is possible. In air-to-air combat it is possible to write 'follow' provided that you are no further away from your opponent than 400mm (40 inches), you are within a 90 degree angle cone subtended backwards from the nose of the target, and your angle of flight is converging, ie, less than 90 degrees with the hostile aircraft flying away from you (see accompanying diagram).

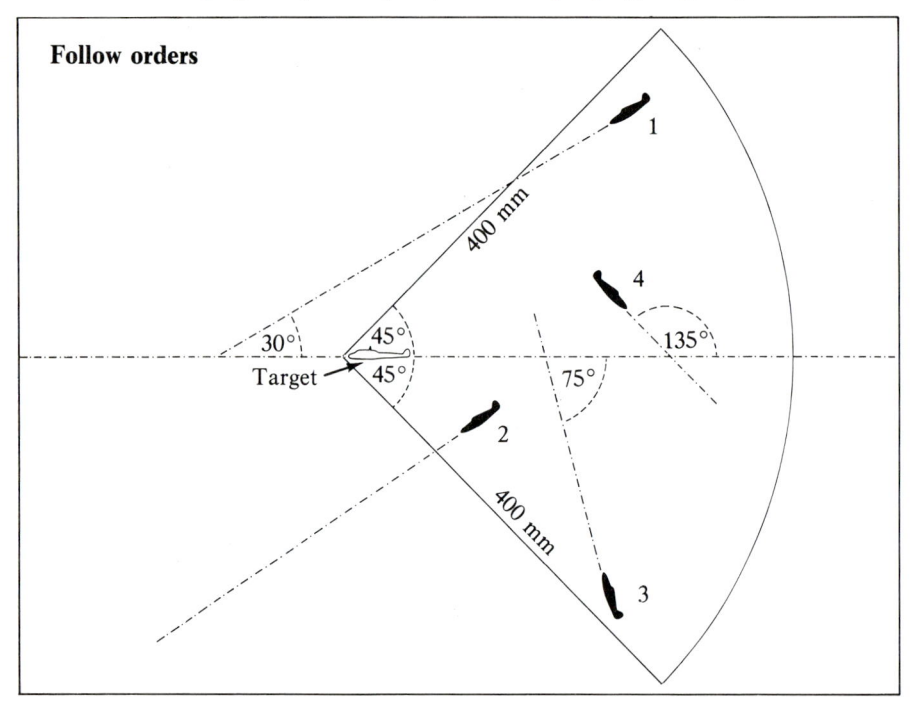

Follow orders

Attackers 1 and 3 are eligible to write a 'follow' order as their course intersects either the course or the track of the target aircraft at less than 90 degrees. Attacker 2 is ineligible as its course does not intersect at all, and Attacker 4 is ineligible as its course intersects the target's track by a greater angle than 90 degrees.

If it is your intention to enter cloud during the move, then you must also write orders for sufficient moves to take you out of it. Failure to comply means that you must straighten out at the point where your written orders end and then keep flying until you do emerge, continuing in a straight line until the end of that move.

7 Move your aircraft, measuring distance as meticulously as possible. If turning, the turning device should be laid in front of your aircraft with the straight edge at right angles to the centre-line of the fuselage and the nose exactly above the required radius of turn. The aircraft should then be moved the desired angle around the circle and end with its nose exactly above the point where the curved

The table-top game

radius line crosses the straight angle line and with its centre-line at 90 degrees to the straight angle line. In the event of a 'follow' order having been written, the move can be taken as described, using the correct turning radius for the speed or any wider radius, to bring you in on the tail of the target aircraft. If your speed is too great and you overshoot, hard cheese! Keep going on the same course.

8 Measure height and re-calculate speed. This should be done even if you think you are in level flight; it is surprising how much you may wander about.

9 Assess potential targets (for fixed armament, the centre-line of the firing aircraft, if extended forward should pass through any part of the target) and announce your intention of firing, your target, and how many seconds of firing. No prior measurement to see if a target is within range is permitted. The firing position is that at the end of the move only.

10 Measure range and deflection, calculate the relative speed of the target and firer, and open fire by rolling the dice once for each second of firing (three seconds maximum), noting the positions of any hits. Range is measured from the nose of the firing aircraft for fixed armament and from the actual gun position for traversing armament to the nearest part of the fuselage of the target. Deflection is the angle between the centre-lines of the aircraft, and comparative speed is calculated as described in Chapter 6.

11 Roll the dice for each hit and resolve damage as described in Chapter 6.

12 Adjust speed for recoil effect where applicable.

13 If, after re-calculating speed at either 8 or 12 above, it is found that speed has dropped down to or below the stalling speed, you have stalled. Speed drops automatically to zero and the aircraft, for the next move, is pivoted about its nose until it is pointing vertically downwards. The throttle *must* be opened at this point or, if this is precluded by engine damage, for this move only it gets its normal acceleration factor to start it moving earthwards, otherwise it would hang there forever like an aerial Flying Dutchman. During this move a stalled aircraft must dice to see if it spins. Whether it spins or not, it continues vertically down during subsequent moves until flying speed is regained. In the case of engine damage the deceleration factor is suspended until a speed of 200 mph is reached. When flying speed is regained, the aircraft is back under normal control unless it is spinning, when dicing to recover from the spin may commence.

14 Recommence the sequence, starting with observation.

A couple of points to be clarified here. If an aircraft is both firing and being fired at, if it is shot down before its own armament scores hits, then these hits are not valid. For example, a three-second burst, scoring a hit with the final second of firing would be cancelled out by a destructive hit on itself in either the first or second second of firing.

For those of you who wish to keep a log-book, the pilot survival probability when shot down is assessed as follows: aircraft explodes—00-05; on fire—00-50; breaks up—00-66; uncontrollable—00-85; forced landing—00-95. If the pilot is seriously wounded, the above probabilities are halved.

Acceleration, deceleration and rate of turn are decided by the speed at which the aircraft is travelling at the commencement of the move.

Scenario suggestions

The most interesting scene is generally a fighter verus fighter game. The simple way is for each player to grab two aircraft, chuck a couple of clouds on to the board at random, roll the dice for height (a score of 82 would mean an entry height of 82 centimetres (nearly 10,000 feet), enter from the opposite sides of the playing surface and get stuck in. If you belong to a club, a hilarious evening can be had with a dozen or more players, each with one aircraft, and an absolute vulgar brawl, the sole survivor being the winner. Either have unlimited ammo and stay in the air, or make it compulsory to come in, land, and spend one complete move stationary on the ground to re-arm after each 12 seconds of firing, then take off again. Ever seen a Gladiator shoot down an Me 262? Nor have I, but I live in hope!

For more serious games I would recommend two aircraft per player initially, four when a degree of practice has been achieved. Try half a dozen Bf 109Es at high level being intercepted by four Hurricanes climbing up underneath with about four clouds beneath the German patrol line. Assume the Hurricanes are being directed from the ground, while the '109s have to rely on visual observation, and cannot leave their patrol height until a sighting is made.

Bomber escorts can be quite good fun. If no-one is willing to fly them, they can be pre-programmed for the bombing run, although I remember one classic case when the '109 escort was taking a terrible beating from Hurricanes, and a Heinkel actually turned back to their assistance! A born optimist. A suggestion here. Last player to arrive flies the bombers!

On a more serious note, keeping a log book for each pilot nationality is worth doing, as with seven different identities everyone is likely to be an ace of one nation or another. Agree on a year of play and a theatre of war, and select aircraft of that year or earlier to use. Don't be too keen to use the Me 262 until you have got some practice, as few players can handle it effectively, although one interesting situation is to set up a horizontal game with three B-17s in tight formation, two '262s trying to get at them and six Mustangs defending. Ignore all considerations of dive and climb and play the game on the level. This game can be kept on the board indefinitely by making all the fighter moves, then moving them *backwards* by the B-17s' move distance leaving the bombers stationary in the centre of the playing area.

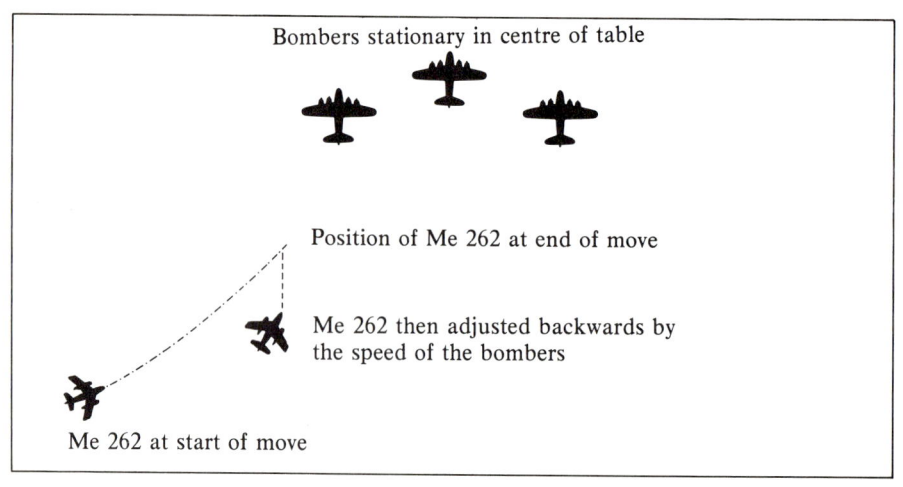

```
            Bombers stationary in centre of table


            Position of Me 262 at end of move

            Me 262 then adjusted backwards by
            the speed of the bombers

Me 262 at start of move
```

There is one particular combat related in *The Fighter Pilots*, by Edward H. Sims, which can be produced as a game, the account of which is of exceptional interest to the student of air fighting, so much so that I have sought the author's permission to quote a large chunk of it in full.

The table-top game 121

The date, September 15 1940. The subject, Sergeant J.H. (Ginger) Lacey, flying a Hurricane of 501 Squadron, code letters SD-F, flying as Pinetree (squadron call-sign) Red Three, from RAF Kenley.

Scrambled to intercept a 50-plus raid coming in between Dungeness and Ramsgate at 'Angels 15'—25,000 feet—501 Squadron failed to gain enough height to intercept the mixed force of Dornier Do 17s and Bf 109s. Lacey, standing on his tail and shooting vertically upwards, stalled and spun off, losing both 5,000 feet of height and the rest of the squadron. Ordered to re-join north of Maidstone, he is all alone, climbing hard, when. . . .

'Then, ahead . . . straight ahead, north-west . . . specks . . . growing rapidly larger. Fighter squadron. Lacey studies the fast approaching bogies . . . single-engined aircraft . . . perhaps 501. They come on. He can see the spinners. Yellow! Me 109s! Coming head-on. Have they seen him? He must act. He's on a collision course with a dozen enemy fighters, each faster than the Hurricane. Instinctively, he pushes the stick handle forwards, the horizon rises and he dives to pass beneath them. But he won't run away. By all the rules of aerial combat he should. He continues his dive, gaining speed, and now the 109s are directly above.

'The rush of air and the roar of the engine increase as speed increases. Now he hauls hard back on the stick handle. He pulls the nose into the straight-up position. This time he has the speed to avoid a stall and the Hurricane rockets upwards. Lacey is now zooming skywards almost directly below the last 109 in the formation, keeping the stick back. The Hurricane, coming over on its back, levels out at the top of a loop directly behind the last enemy fighter. Lacey is upside down, held firmly by seat straps, eyes fixed on the Messerschmitt ahead. 150 yards! Perfect firing range. His loop was timed to perfection. He has never shot at another aircraft on his back. The gunsight is adjusted for a normal drop in the trajectory of shells from an upright position. He will have to allow for this and fire well above the 109! He must act, speed is falling. The 109s are cruising homeward at about 240 mph and though he has built up speed in his dive, when he loses that momentum the 109s can pull away. So far they haven't seen the audacious approach. He lines up the enemy wingspan in the glass.

'Now. Fire! His thumb (pointing earthwards) pushes the button. Eight machine-guns spit a stream of armour-piercing and incendiary shells. Lacey, head down, watches the effect. The roar of his guns and the vibration of the aircraft add to the strange sensation of attacking on his back. But he has compensated accurately. The 109 staggers from the blast of close, concentrated fire. Incendiaries crash into the engine and into the fuel tank behind the pilot and in seconds a black stream of smoke pours out. Complete tactical surprise! Now the 109 turns out of formation, falls off to the side and bursts into flame. It plunges straight down.

'Finally, Lacey rolls over. The blood drains from his head and he sees more clearly the 11 dark enemy fighters, crosses on their wings directly ahead. Still they haven't seen him—or the destruction of their comrade in the rear! He manoeuvres in behind a 109 to the left . . . distance 250 yards, quickly he centres rudder and stick for proper attitude and lines him up in the sighting glass. Wingspan almost bar to bar, squarely in the orange circle. The firing button! For the third time the wing guns belch more than a hundred shells a second. Though the pattern of fire is not fully concentrated at 250 yards, shells strike the second Me 109 instantly, tearing pieces out of the fuselage and wings. Lacey holds the button down. More shells find the mark. The second victim now begins to stream white smoke. Lacey knows he's hit the coolant. The 109 is doomed, for without coolant the Daimler-Benz will overheat.

'Lacey takes his finger off the firing button. The pilot will have to jump—over England or the Channel. Suddenly anti-aircraft fire begins to dot the sky. It bursts uncomfortably close. And at this moment the remaining ten enemy fighters, at last aware of an impudent intruder do what they should have done in the beginning. They split into two groups, half turning right and half turning left. They are coming around to get behind him. If he turns after one group, the other will curve in behind. His ammunition is low. They have a speed advantage. He is certain to be caught from the rear by one half or the other! Yet he still has a few rounds and one 109 in the group banking left is trailing behind. Perhaps he just has time to get him and still get away. It's a gamble. Left rudder, left stick. Lacey leans into a hard left bank to line up behind him. The group to his right banks more sharply to come around behind. No doubt the enemy pilots can clearly see red and blue circles on top of the turning Hurricane's wings and no doubt they admire the tenacity of this interceptor, who should be running away from the battle—who should not have initiated it at all.

'Lacey keeps stick handle back, standing on his port wing, and lines up behind the last 109. He's at maximum range. Time is precious. Fire! For a second the Hurricane's eight guns lash out in a staccato. The firing becomes erratic. One gun empties after the other. All silent. No more ammunition! He sees white streaks ahead, in front, on both sides. He's in the trap. The 109s which banked right are behind. Dive! Instinctively his right hand pushes the stick handle forward and he almost lunges off the seat as the fighter's nose dips and dives for earth'.

Lacey dived into a friendly-looking cloud, losing his pursuers in the process, and went home on the deck, totally unscathed.

This was, of course, not a typical sortie of the period. Solo attacks at odds of 12 to one were beyond both the ability and temerity of the average pilot. It should be stressed that Sergeant (later Squadron Leader) Lacey was an exceptional aircraft handler and a first-class marksman. If you care to reproduce this sortie in a game you will be lucky to get away with it, even if the number of 109s is reduced to four.

So far I have made suggestions incorporating the better-known aircraft. Try not to neglect the more unfamiliar theatres of war as the very novelty of these can be rewarding. Biplanes against monoplanes make a change.

For those of us who have no regular opponent, a few suggestions for the solo game, and if you keep a log book, you can build up quite a respectable score. The requirement of a solo game is that the response of your opponent is automatic. This is limiting, but certain opportunities do arise.

The first game is as a bomber destroyer. Set up half a dozen or so B-17s in close formation and have a go at them in a FW 190A-8, which is a heavily armoured bomber destroyer. Play the game in the horizontal plane rather than the vertical, and instead of moving the bombers, after making your move, pull your aircraft rearward by the bomber's move distance as described before. You will have to operate the bomber's defensive fire, but see how many sorties you can survive.

The second game is the tank-destroyer. This is played in the orthodox manner as described in Chapter 7. You will have to do the ground-to-air firing as well. A word of warning. Do not make the flak too dense or you won't survive the first attack.

Dive-bombing is the third suggestion and is merely an extension of tank-busting.

Last but by no means least, try a torpedo or skip bombing attack. This is played in the horizontal plane and is described fully in Chapter 7.

Chapter 9

Large-scale air warfare

The difficult thing to learn about reproducing large air battles and campaigns is how far to take them, and the problem is always deciding what to leave out. This is, of course, very much a matter of taste, as is the scale to which one is willing to operate. It is also necessary to lay down clear victory conditions, which may well vary according to the objectives of the respective opponents; the larger the scale and the more detail incorporated, the more difficult this becomes.

Another problem peculiar to the reproduction of air warfare also exists. The basic skirmish game previously described is only suitable for handling relatively small numbers of aircraft, and an encounter between even such small forces as 50 per side would flood the playing area totally. Also, air warfare, by its very nature, prevents the effective up-grading of models, so that where one soldier figure may represent 20 or so men in the Napoleonic era without undue loss of realism, one aircraft model cannot represent anything other than one aircraft. When fighting a land or sea battle, the player is in the position of the commanding general or admiral. In the air, there is no commander up with the action to plug holes, launch attacks, move up reserves, and generally respond to the tactical needs of the movement. It was tried on a number of occasions but generally failed, not least because in the air things happened in split seconds rather than minutes, and no one man could be aware of the whole situation. Command in the air was generally at a relatively junior level, and before anyone says 'what about the wing leader?' let him stop and consider that in a fairly typical circus operation in 1941, no less than seven wings could be used, each one with a specific role, and each with its own wing leader.

The real commander in an air battle is, of course, the ground controller or, where no control is possible, the raid planner. What we have to do then, is to set up our own operations room complete with plotting table, about which we shall elaborate further shortly. Quite a few board games simulate air campaigns, but these suffer from two main faults. The first is that casualties are assessed in terms of a whole day's, or even a week's fighting. It is rather dismaying to have eight or ten squadrons totally wiped out in one round, even though they are replaced during the next. Also, the hexagonal grid playing area, although excellent in many ways, is simply not precise enough for exact interception, and interceptions were often missed by as little as a thousand feet or a few seconds. But before proceeding further, let us examine a few middling to large air operations and sort out what the requirements really are.

France 1940 was bit of a shambles. Not much ground control was evident; a high proportion of air fighting was in small numbers and due to chance encounters, and any attempt to simulate this would end up more like a game of ludo. The Battle of Britain in 1940 was better organised. Large scale German raids were meticulously planned to achieve certain objectives and, weather and errors by junior commanders on the spot excepted, they were carried out in a most determined manner. The targets were specific; convoys in the Channel, airfields, radar stations, aircraft factories, and finally London. The German player will be the raid planner, and can plan his strike(s) to the finest detail. Provided we allow for the odd banana skin in the form of weather and navigational errors, his game can be pre-planned from the outset and played in strict accordance with the plan. If we are talking in terms of one large raid or a multiple series of raids, his success can be judged by the number of bombers successfully raiding the target(s) in proportion to his losses. The British player's success can also be judged by this yardstick. His task is to intercept all German raids and inflict unacceptable losses on the raiders. He suffers, however, from the disadvantage of having to wait for the attack to develop, and of not knowing which attacks are feints and which are real. He does have the advantage of radar, and of the Observer Corps, so that he always has a fair picture of what is happening, even though at times he will be unable to do much about it.

In 1941 were instituted the 'Circus' operations. These consisted of bait in the form of half a dozen or so bombers, escorted by up to 200 fighters. The bombers would attack a Luftwaffe airfield; Abbeville, St Omer, or wherever, but the object of the exercise was to draw the German fighters into a battle of attrition. The victory conditions for a Circus could be assessed by the comparative percentage losses on each side.

In 1942, the Americans started their daylight raids which, although limited at first to fairly short incursions, by 1944 were penetrating deep into Germany with a massive long-range fighter escort. The trouble with simulating these raids is that you need so much information as to German strengths and dispositions, also a simply colossal playing area. I would suggest that these are simply too large to be worth attempting when you can get an equal amount of enjoyment from something on a smaller scale.

Also in 1942, on the other side of the world, the big carrier battles had begun. These are in many ways excellent subjects for simulation, the number of aircraft per side being restricted to a few hundred, and it is also very easy to determine when a base is knocked out. It sinks! Also the types of aircraft involved are limited to a mere handful, which reduces complications somewhat. The problem is to produce an acceptable simulation of air reconnaissance.

On the Russian front, the air war was fought almost entirely in support of the land forces, and it is difficult, if not impossible, to disentangle the air war from the ground operations, as is also the case in the Western Desert, Italy and Normandy, and large-scale ground support campaigns are rather outside the scope of this chapter, although the air supremacy role could be effectively simulated and the interdiction results assessed in accordance with its success or otherwise.

Finally, the rabbit out of the hat. Malta! If you are campaign minded, Malta is ideal. The numbers are not excessive, the distances and timings are short; no question arises of it getting involved in a land battle as the front line moves, and both sides can be offensive, so that one player is not committed to the defensive for the entire game, as happens in most other situations. Also, both sides had radar.

Malta could not be ignored by the Axis as it was an offensive base for cutting their supply lines to the Middle East. It could not be ceded by the Allies for the same

Large-scale air warfare 125

reason. Malta was critical to the entire North African land campaign. But more of Malta later.

What are the requirements for setting up an operations room table type playing area? The first consideration is scale. Contemporary photographs show that the real thing was quite a size. As few of us have much room to spare we have to settle for something smaller, but not so small that accuracy is lost, while an essential requirement is to be able to reach over and make accurate measurements in the centre of the table without knocking everything flying (no pun intended). After some trial and error I settled for one centimetre to one mile. This, when set up on a playing surface two metres long by one and half metres wide, gives a usable scale area of 200 by 150 miles. If we take the Battle of Britain as a worked example, this will give nearly all the 11 Group area, and if you make the best use of it, a fair lump of France also.

The way to do this is to take a straight line just to the north-east of Middle Wallop, Andover, Northolt, North Weald and Martlesham as the top edge of the board. The left-hand side of the board is taken down at right angles, passing through the Needles just west of the Isle of Wight, and it should just cross the French coast about 30 miles south-west of Le Havre. The opposite side will come down into Belgium about 25 miles north-east of Lille. Do not be put off by having to prepare a map. Absolute accuracy is not essential and anywhere within five miles will do, as both you and your opponent will be using the same air space. I wouldn't quite suggest that you try it freehand, but it doesn't have to be a shining example of the cartographer's art. Any cheap road map of southern England will do to work from as all you need is the outline of the coast, the airfields dotted in and named, and a few of the major towns, Winchester, Aldershot, Canterbury, etc, put in as navigational aids. For France I would suggest the RAC Continental Handbook as a guide.

Having prepared our map, criss-cross it with a grid of faint lines at 200 mm centres. This is a useful navigational aid, would you believe? The area we have chosen is not ideal, as it omits the Debden Sector entirely, the Duxford Wing from 12 Group, about half of the Northolt Sector, and touches just one corner of 10 Group. The Cherbourg Peninsula is outside its scope, as are most of the Luftwaffe bomber fields in France. If you want to do the entire Battle of Britain, get yourself a larger board and rope in some assistance. You'll need it. The board we are using does, however, have one major advantage in that it is almost entirely within the British high level radar cover, so that any high level (ie, 15,000 feet plus) raids coming in from the base line can be plotted at once. It also gives a representative slice of the action without becoming too large to be easily handled. The bomber fields behind the baseline are not critical, as the bombers can afford to take off and spend 30 to 45 minutes gaining height and sorting out their formation. The critical part is, of course, getting the fighter escort with them without wasting time and fuel as the escorts had a very short range indeed and their time on station was extremely limited.

Carrier battles have one inestimable advantage. They do not need maps, unless you like to chuck a small island such as Midway into your scenario. Be careful if you do though; you might get side-tracked into a complete Japanese invasion, which is not the object of the exercise. It may have been in 1942, but all I want to give you is a carrier battle. All you need to do is to insert the 200 mm grid, and agree which direction is north. The Japanese enter from the north or east; the Americans from the south or west. What could be simpler?

Malta is very easily covered. With the 150-mile base line at the bottom of the board and the island itself given about 25 miles sea-room to the south, at least seven eighths of Sicily can be accommodated, including most of the northern coastline, only Marsala and Trapani being omitted.

We shall be returning to all these scenarios shortly in greater detail; what we must first establish before so doing is a method of covering all necessary operations regardless of theatre.

Our first concern is weather. If we are doing a large-scale operation, weather unfit for flying would lead to the operation being cancelled, so this need not concern us. It must, however, be considered in a campaign, where it could, for example, give a hard-pressed defender a respite for a few days, or alternatively turn a grass strip into a bog which persists even after the weather has improved. What we need then are two charts, one to determine non-flying weather, the other to sort out the actual degree of weather when fit for flying. On top of this we must legislate for changeable weather, also for wind speed and direction. To give a fairly extreme example of why this is necessary, if a bomber took off having been briefed to expect a cross-wind of 30 mph at right-angles to its course, and the predicted wind was wrong, and instead it was blowing at 30 mph directly against the bomber (and worse cases occurred in practice); unless a landmark could be identified on the ground, or a star-sight taken (not always so easy; the last time I came back from Munich we were between two solid layers of cloud from five minutes after take-off right to the English coast); then dead reckoning, the practice of calculating the theoretical position, would produce in a mere two hours' flying at 200 mph an error of 80-odd miles. Always give the enemy the opportunity of getting lost.

Anyway, back to the weather. Non-flying weather can be defined as snow, fog, or very low cloud, which really I suppose can count as a sort of fog. Rain in itself does not generally prevent flying, except when enough of it makes airfields unusable. Anyway, for a campaign, use the following chart which is coded as follows: S = snow in Western Europe and Russia, sandstorms in the desert, and storms in the Pacific; F = fog; and W = flyable weather.

	Western Europe			Middle East			Russia			Pacific		
	S	F	W	S	F	W	S	F	W	S	F	W
Jan	00-10	11-25	26-99	00-05	06-10	11-99	00-20	21-30	31-99	00-10	—	11-99
Feb	00-10	11-25	26-99	00-05	06-10	11-99	00-20	21-30	31-99	00-10	—	11-99
Mar	00-05	06-20	21-99	00-05	—	06-99	00-15	16-25	26-99	00-05	—	06-99
Apr	—	00-05	06-99	00-05	—	06-99	00-05	06-15	16-99	00-01	—	02-99
May	—	—	00-99	00-05	—	06-99	—	00-05	06-99	00-01	—	02-99
Jun	—	—	00-99	00-05	—	06-99	—	—	00-99	00-01	—	02-99
Jul	—	—	00-99	00-05	—	06-99	—	—	00-99	00-01	—	02-99
Aug	—	—	00-99	00-05	—	06-99	—	—	00-99	00-01	—	02-99
Sep	—	—	00-99	00-05	—	06-99	—	—	00-99	00-01	—	02-99
Oct	—	00-05	06-99	00-05	—	06-99	00-05	06-15	00-99	00-05	—	06-99
Nov	—	00-05	06-99	00-05	06-10	11-99	00-10	11-20	00-99	00-05	—	06-99
Dec	00-05	06-20	21-99	00-05	06-10	11-99	00-20	21-30	00-99	00-10	—	11-99

Having thrown the dice and referred to the weather chart for the basic weather, further adjustments must be made as follows. Fog, snow, sandstorms and other storms all cancel flying. Flyable weather permits flying, but this tends to be changeable, especially in Western Europe, and we must legislate for this. Assuming that we have flyable weather for the day, roll the dice yet again. The

Large-scale air warfare 127

number on the red die is the cloud cover in tenths. Perhaps I had better explain this in more detail.

If the sky is completely overcast, then it is described as ten tenths cloud. If half the sky is clouded over, it is then described as five tenths cloud. In other words, the proportion of sky covered in cloud is assessed in tenths. The number on the black die must be doubled; this then gives the height of the cloud base in thousands of feet. A throw of four on the red die and six on the black gives a situation where cloud cover is four-tenths at 12,000 feet. The amount of cloud determines whether or not it is raining. Halve the cloud cover, multiply by ten, and this is the probability of rain. For example nine tenths cloud, halved, then multiplied by ten, gives a 45 per cent probability of rain. Use this only down to five-tenths, as anything under this will only give showers, which are not really crucial. Cloud cover and rain should be checked on an hourly basis. 12 hours of rain in any three-day period should make a grass field unserviceable. Airfields with tarmac or concrete runways are generally impervious, although not always. For example, Takali on Malta was always susceptible to flooding. Advanced airfields, with Sommerfeld track laid on the ground, are in between these two extremes, and should be diced for on a 50:50 basis. A flooded airfield can be assumed to take two whole rainless days to become serviceable again.

Snow is assumed to fall fairly evenly all day. The runways and taxi-strips are assumed unserviceable during the fall, and snow-clearing can commence at daybreak on the following day. Assume three hours to clear a day's snowfall, increasing in proportion to the square of the number of consecutive days on which snow actually fell. An example of this might be six days' snowfall followed by a two-day break, followed by three more days of snow. A bit extreme, but then, the Russian winter is extreme. Ask Napoleon! The square of the six days' snowfall is 36, multiplied by three hours = 108 hours clearing time. The two-day break of 40 hours (daybreak—08.00 hours) is then spent in partial clearance, but when the snow commences to fall again on the ninth day, 68 hours work remain to be done. The square of the next three days is nine multiplied by three hours = 27 hours. 27 + 68 = 95 hours' work to get the airfield operational once more. Starting from dawn on the 12th day, the airfield is out of action until 07.00 hours on the 16th day.

Storms in the Pacific leave a heavy swell the following day which doubles landing and take-off times.

We must now consider plotting and tracking. A considerable amount of research would give us the plotting capability of any nation. I would suggest, however, that this is an unnecessary complication and if we can standardise on a method it will not only save time but also a considerable amount of effort, as the player will not have to get to grips with half a dozen different systems with their inherent strengths and weaknesses. None of them were very exact anyway. The system I would propose is the British system of 1940. The coast was ringed with radar stations which looked out to sea. These stations were of two different types; the 'Chain Home' main radar, and the 'Chain Home Low' for tracking low-level raids. Both were inaccurate as to height, also in assessing the number of aircraft in a formation. The combined detection capability of both types can be assessed as in the table at the top of page 128.

These figures are, of course, variable, but will serve as a reasonable approximation. 120 miles is the maximum range regardless of height, and at the other end of the scale radar is virtually useless at detecting aircraft flying at less than 500 feet.

Minimum altitude for detection (feet)	Range (miles)	Minimum altitude for detection (feet)	Range (miles)
500	18	8,000	76
1,000	22	9,000	83
1,500	28	10,000	90
2,000	35	11,000	96
3,000	42	12,000	102
4,000	49	13,000	108
5,000	56	14,000	114
6,000	63	15,000 plus	120
7,000	69		

At the back of the radar stations was the Observer Corps, spaced out at five-mile intervals and equipped with telephones and some odd-looking optical aids. As a raid crossed the coast and vanished behind the radar screen, the Observer Corps took up the tracking visually, continually 'phoning in reports. The information from the OC posts and the radar stations was fed back to the Fighter Command Filter Room at Stanmore, where it was processed and cross-checked, appearing after a delay of about five minutes on the plotting table, which is what we are at such pains to produce.

Having no actual radar stations and observation posts, we are dependent on the enemy player to fulfil this function for us, and position his own plots accurately on the table, commencing at take-off. However, the defender cannot react until the plot reaches the offshore range at which the radar would pick it up. The attacker may also tell great whoppers about the size of his forces and the height at which they are flying. This is to compensate for the inaccuracy of the radar. The whoppers are limited to up to 20 per cent on numbers and 20 per cent on height; there is, of course, nothing to stop him telling the truth if he thinks it will mislead the enemy. These terminological inexactitudes change, however, as the raids come within visual range (ie, five miles offshore). When observed, the exact number and composition of the raid is known to the defender, and the height is given to within ten per cent of the actual. When observed from the air, of course, everything is exactly given. The one exception is when cloud obscures ground-to-air observation. If the raid is flying above the cloud base, seven tenths cloud means only a three tenths (30 per cent) chance of a sighting per move. If no sighting is obtained aural tracking will continue, but the original information given on height and numbers will be perpetuated until a sighting can be made.

Very low level raids will be first plotted by visual means only, at five miles off the coast, and will be tracked through by the Observer Corps in the normal way. Advantage cannot be taken of cloud cover by these raids as zero feet blind flying is out of the question.

Our next need is counters to represent aircraft units. You can, if you wish, filch counters from a board game, but I do not find this satisfactory. The best method is to mount a small piece of plastic card vertically on a base, just large enough to pencil bits of information on as the game proceeds. The base should not exceed one centimetre square, and should have a small arrow on it to indicate the exact position of the formation leader, also the precise direction of flight. This base needs to be weighted in order to prevent it tipping over in moments of excitement. It is advisable to have plenty of spares available. Remember that a *Geschwader* can split in the twinkling of an eye into nine *Staffeln*.

Our next concern is to set up the order of battle. This may be done on either a

Large-scale air warfare 129

historical or fictional basis, according to choice, although the only purpose served by an entirely fictional scenario would be in the event of a war situation played 'blind', with neither side having any intelligence about the strengths and dispositions of the other, and having to produce targets as a result of reconnaissance. This has a certain attraction but is not generally as interesting as the recreation of an actual time and place, using the correct forces and dispositions.

We next need to examine game requirements and how aircraft performance will relate to the game. Time is the first thing to be considered. The game move should be not too short, and not too long. Trial and error has produced a six-minute move length, although it is often necessary to work in half moves if an interception is imminent. If six minutes seems an odd time, consider that dividing the speed in mph by ten gives the move distance in miles. Half-moves are necessary for precise interceptions, but orders from the ground to units in the air can only be given once every full move.

Aircraft performance data needs fairly careful consideration. Take-off can be standardised for bomber units of varying sizes and also for fighters. We can say that at the end of the first move, a unit of up to 12 fighters can take off, form up at a height of 1,000 feet, ready to commence their climb, and that bombers take two moves to achieve this. These times should not be shortened for smaller formations. Where more than one unit is taking off from the same airfield to fly in formation, these times should be adjusted pro-rata. This may seem severe, but the extra time needed to get two formations into formation with each other would be about this. The large American bomber formations used to take a tremendous amount of time to get into formation, even though taking off from many different fields. For example, a 36 aircraft Bomb Group could take 75 minutes to form up and reach 8,000 feet. By the time two more Bomb Groups had joined up to form a Combat Bomb Wing, and a further two Combat Bomb Wings had tagged on at the coast at 14,000 feet to form an Air Division, nearly two hours would have passed. Even this was not the end however, as a final rendezvous of three Air Divisions was possible off the Dutch coast at 20,000 feet, two and a half hours after the first aircraft took off. This I grant is a very extreme case, but do I make my point? Finally, before we leave the subject, carrier aircraft take half as long again.

Our next consideration is rate of climb. This has been dealt with in the skirmish game in fair detail; just a little more elaboration is required. As we have to plot speed over the ground, a rule of thumb is needed. If a fighter is climbing hard, we can assume that its speed over the ground is 45 per cent of its maximum level speed, ie, a maximum level speed of 360 mph gives a ground speed of 160 mph in the climb. If distance is not required, and the fighter is simply making height, it can be assumed to be in a spiral climb. This means that on-table movement is not required; only the altitude changes. For laden bombers with their much slower rate of climb, we can say that their forward speed while climbing is half their maximum level speed.

This brings us on to the whole question of speeds and what is very important, although not touched on in the skirmish game, range. Speeds above maximum level cannot be considered in the campaign game as, although attainable in a dive, they were never held for long enough to affect us. The maximum move distance for all aircraft is thus limited by their maximum speed in level flight.

Aircraft formations, unless in a terrible hurry to either escape or intercept, will use their crusing speed. Where there are two types of aircraft in the formation with differing cruising speeds, the formation speed will be that of the slowest. Bombers are unable to use their maximum speed until they have unloaded, and only then if they are prepared to lose 2,000 feet of height per game move on the way home, and hold their course in a straight line.

We must now consider range, as with most single-seater fighters, it is critical. Maximum ranges quoted in published performance figures are usually a theoretical calculation only. They do not allow for fuel used in warming up the engine, in take-off, in any full throttle combat, or any safety margin by way of a reserve. Occasionally the practical combat radius is stated, but this also needs watching as, while it covers many of the essentials, it does not cover exceptional situations such as an abnormally high operating altitude, or an exceptionally long time spent at full throttle in a combat situation. Range then is, in fact, controlled by fuel consumption. Let us take an example and analyse it.

The Bf 109E in 1940 is as good a place to start as any. The maximum range is stated as being 413 miles. Operating from airfields on the Pas de Calais it could operate over North London, an approximate distance of 100 miles or a 200 mile round trip making no allowance for diversions along the way. We know that its endurance was about ninety minutes, give or take a bit, which would give an average speed of about 133 mph. We also know that quite a few ended up either in the Channel or on the beaches of France through running out of fuel. So what are we able to deduce? First that 90 minutes gives us 15 game moves. Secondly that one complete climbing move starting from 1,000 feet would take us up to somewhere near 17,000 feet, which is a reasonable operating altitude (see Chapter 8 for rate of climb). Also that the combat radius is barely 25 per cent of the range. Not much to go on so far, is it? Now we must begin to make some arbitrary assumptions, basing everything on consumption of fuel. After all, a '109 if unopposed and flying at economical cruising speed should be able to get a lot further inland than North London and still be able to return.

First we ignore the warm-up period as irrelevant. Assume that combat cruising speed is flown in all general situations, and this consumes one unit of juice per six-minute move. Assume that in all other situations, such as take-off, climb, combat and escape, full throttle is needed which uses fuel at twice this rate. Then finally assume that one move of take-off, one move of climb, two moves of combat and one of escape, plus 11 normal moves at combat cruise make up the normal duration giving no margin for error. This gives us 21 units of fuel, or two hours six minutes' flying time at relatively low height and speed. Let us see how this works out in practice. Move 1, six minutes, take-off and form up. Move 2, 12 minutes, climb to 15,000 feet, four fuel units used, 14.2 miles covered. Move 3, 18 minutes, combat cruise, five fuel units now used, 44.6 miles now covered. Move 4, 24 minutes, combat cruise, six fuel units now used, 75 miles now covered. Move 5, combat cruise again, seven fuel units now used, 105.4 miles now covered and we are over London, with 14 fuel units in hand, and needing only four to take us home again. Plenty, even if we have to fight our way out.

The example just given is fine for fighters engaged in a *frei-jagd* or free hunt. But what if they have to escort bombers? Back to Move 2, lads! Here we are then, at 15,000 feet, four units used, only three miles covered, and where the hell are the bombers? Here they come, late as usual. Up into the top-cover position 3,000 feet above them. Half a move wasted. Dozy lot of Dorniers! Only eight miles covered by the end of the move. Move 3, combat cruise, five fuel units now used, 21.6 miles now covered. Yes, I know we are on combat cruise and are travelling much faster

Large-scale air warfare 131

than the Dorniers. It's just that we zig-zag to keep level with them. We travel much further through the air, but only the same distance over the ground. We have to keep our speed up in case we are bounced. At low speed, we would be sitting ducks. Move 4, combat cruise again, six fuel units now used, 40.2 miles now covered. On to Move 7, nine fuel units and 96 miles covered. We will be over the target halfway through the next move, then turning to go . . . 'Achtung, Schpitfeuer!' No it's not, it's Hurricanes. Never mind. Move 8—full throttle in combat, 11 fuel units used, staying with the bombers, 100 miles to go home. Move 9, still in combat, 13 fuel units used, still escorting the Dorniers, 81.4 miles to go. Move 10, here come those last 50 Spitfires that our Reichsmarschall keeps telling us about. 15 fuel units used, six to go, and still 61.8 miles to cover. Good job we weren't intercepted until the final run to the target. Move 11, and still we stay with the bombers. 17 fuel units used, four to go, and 43.2 miles back to the chateau. Move 12, time to go lads. Abandon the bombers and belt off home at full throttle. Only two fuel units left, but only 7.2 miles back to base. Just enough, as if we arrived back over our airfield on our last unit, each fighter would have stood a 20 per cent probability of running out of fuel in the final circuit and ending in the local midden with severe damage.

As you may deduce from this little tale, if the bombers had been intercepted on the way in, so much fuel would hae been used by the escorting fighters that they would have had to abandon the bombers much earlier. Further points which arise from this sequence are that when fighters are escorting bombers, even though under attack they keep drifting along with the bomber stream. The dogfight situation persists even though the escorts are out of ammunition. If the escorts had been detached to fight off an attack, the dogfight would have taken place in the same area as the interception, thus using more fuel without getting any nearer home. To break off a dogfight you must be in a faster aircraft, or your opponent must be either willing to let you go, or unable to stop you, the most usual reason for this being that he is also out of ammunition. The escape move must always be at full throttle.

A brief point here. After two full moves of combat, a unit is assumed to be out of ammunition. When you only have a few seconds firing time, 12 minutes is more than enough to shoot off the lot. Anyway, the example works, so from it we can evolve a range formula to apply to all fighters. We earlier calculated that the average speed of the Bf 109E in a combat scenario was 133 mph, which is the maximum speed divided by 2.67. Combat radius can generally be considered to be 25 per cent of the maximum range for land-based aircraft, and 20 per cent for carrier aircraft. From this we can state a rule of thumb for fuel units which can be calculated as follows: $10\left(\dfrac{\text{max range} \div 2}{\text{max speed} \div 2.67}\right) + 5$.

This will apply to fighters and carrier based bombers only. For land based bombers it should be sufficient to plan their trip not to exceed one quarter of their stated range. I have prepared a fuel unit chart for the main types which is given hereunder, although no allowance has been made for external fuel tanks. Always remember that fuel units are a measure of endurance rather than range.

Should you wish to hang auxiliary tanks on your aircraft to increase the range, you will r+5. ɔ do some research and calculations yourself. You will need to know the basic amount of fuel carried internally, and relate this to the fuel units. Be careful, because an Imperial gallon is larger than a US gallon. Having done this, you will be able to calculate how many extra fuel units you have in the drop tanks.

Finally, make a ten per cent reduction in the total number of fuel units to compensate for the increased weight and drag.

Fuel units	Aircraft types	Fuel units	Aircraft types
17	Polikarpov I-16.	33	Ju 88C-4, P-40C Warhawk.
18	Me 262a, Fiat G 50.	34	P-39D Airacobra, P-51B Mustang.
19	Spitfire IX, Spitfire XIV, Yak 3, Ki 61 Hien.	35	Fiat CR 32.
		36	Yak 9D.
20	Bf 109F, Macchi C 200, LaGG 3, Yak 1M.	39	F4F3 Wildcat, F4U1 Corsair.
		40	Nakajima B5N2.
21	Bf 109E, Bf 109G, FW 190A-8, FW 190D-9, La 5 FN.	43	P-38J Lightning, Barracuda II.
		44	F6F3 Hellcat, N1K2J Shiden 21.
22	Spitfire VB, Typhoon IB, FW 109A-3, Macchi C202, MiG 3.	45	Skua.
		47	Ki 43-I-2B Hayabusa.
23	Bf 110C-4.	49	Ki 43-Ic Hayabusa, A6M6C Model 53 Zero.
24	Spitfire I, Hurricanes I and IIc.		
25	P-47D Thunderbolt.	51	Mosquito VI.
26	Ki 44 Shoki.	53	Nakajima B6N1.
27	Gladiator, Morane 406, Dewoitine 520.	54	A6M2 Model 21 Zero, D4Y Suisei.
		56	Aichi D3A1.
28	Tempest.	58	SB 2C1 Helldiver, Swordfish.
29	SBD Dauntless.	63	TBM 1C Avenger.
30	Fiat CR 42.	64	Beaufighter VIF.
32	Ki 84 Hayate.		

The next step for anyone wishing to mount a raid, is to plan it. This has to be done in detail. Take-off and rendezvous times for both fighters and bombers must be calculated, turning points and targets clearly stated. The forecast wind must be taken into consideration here. It is not necessary for the raiding player to commit all his force at once; the defender has to husband his resources against a further strike which may or may not materialise. He daren't be caught on the ground.

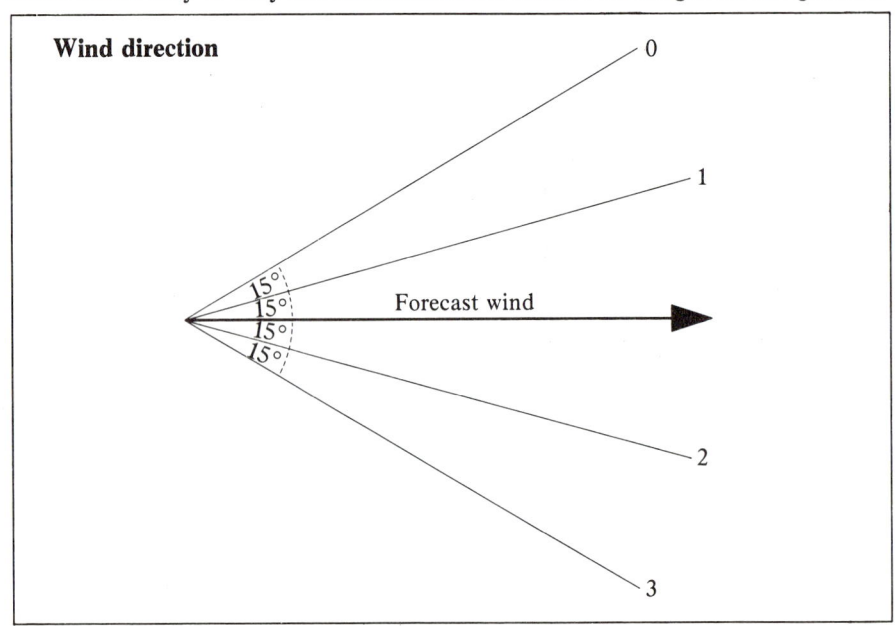

Wind direction

Forecast wind

0
1
15°
15°
15°
15°
2
3

Large-scale air warfare 133

The forecast wind I would suggest can profitably be 20 mph from the south-west in all scenarios. As the attacker takes off, the defender rolls a normal die. A throw of six means that the forecast wind is wrong. In this event, roll the percentage dice. The number on the red die will give the new wind direction, as shown in the wind direction diagram. The number on the black die will be the change in wind speed, in mph, which can be added to or subtracted from the original wind speed at the defender's option. For example, a six on the black die represents 6 mph which, added to or subtracted from the original wind speed, can be adjusted to either 14 or 26 mph at the defender's choice. This is because weather changes almost invariably favour the defence. 0 on the die = 10 mph. The wind is checked in this manner every ten game moves, although in the event of a second change of wind direction, the change of direction is set up on the last wind direction rather than the original forecast wind. Cloud cover is decided by dice as previously laid down, before raid orders are written, and does not change during the game.

When the wind changes, it blows aircraft off course. Until they notice this, they have to continue on their original compass course. To be able to observe this, they must get a landmark on the ground to check their position. This is automatic if they are flying under the cloud base and over land, but if they are above cloud or over the sea it is not so easy. The chance of an air-to-ground pin-point is based on the cloud cover in the same manner as the ground-to-air sighting given earlier. If, however, an air-to-ground pin-point is not obtained, the aircraft will continue to be blown off course. This is calculated with a triangle of forces and shown in the accompanying diagram, in which we can see what happens to an aeroplane in this situation.

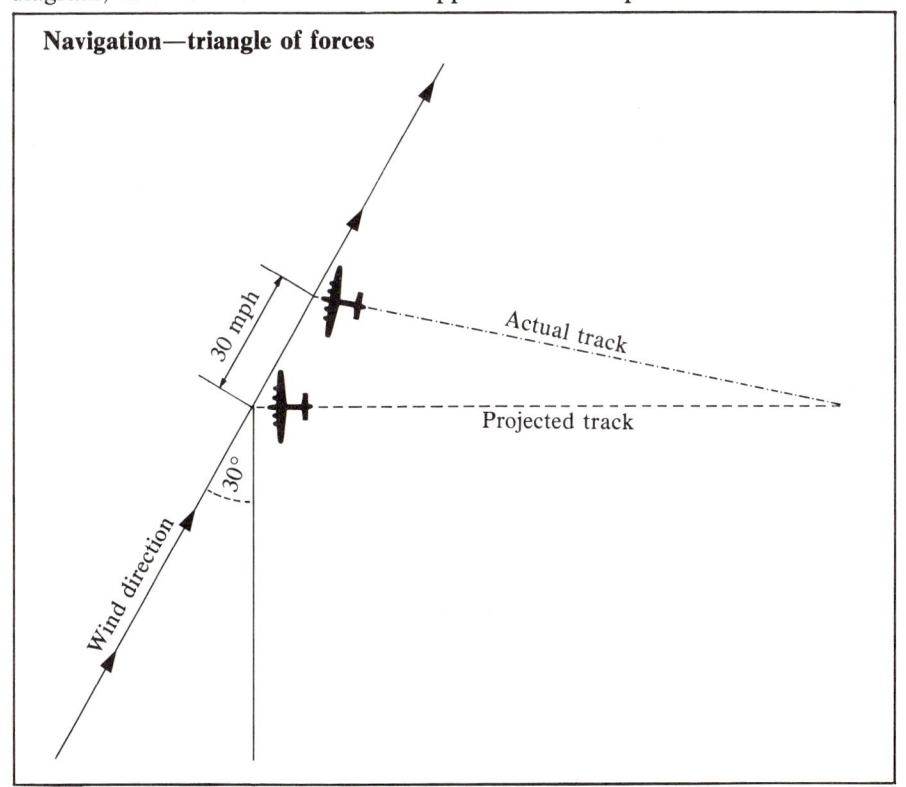

Navigation—triangle of forces

You could, in fact, simplify the planning if you so wish, and use a wind of nil mph from the south-west as a forecast wind and simply plot any changes. Personally I prefer to do it the awkward (and correct) way. If the wind has changed and gone unnoticed the aircraft proceeds on a straight heading, going ever more adrift. If a calculated turning point comes up when an aircraft is off course, it makes the turn in the exact number of degrees that it would have had the wind not altered. When the wind change is noticed, the aircraft turns back on course for either the target or the turning point, whichever is most sensible. Having picked up the wind changes, it can now carry on as normal. One final point. Defenders are not affected at all by wind changes.

We now at last turn our attention to combat. We need not consider the respective merits of individual types of aircraft in any great detail, as what counted more than anything was positioning for the initial strike. Combat can be divided into two main types, the interception which, if surprise is achieved, is potentially very damaging, and the confused whirling mêlée, or vulgar brawl, in which lots of shooting is done but few hits achieved. Let us examine both types step by step, starting with the interception move.

Intercepting fighters, being controlled from the ground, will always manage to spot their target unless it is screened by cloud, in which case previous observation rules apply. The target force, without guidance on where to look, has only half the chance of observing an intercepting force, ie, 50 per cent at best, or in the event of, say, six tenths cloud cover, the intercepting force will have a 40 per cent chance of spotting the raiders; the raiders only a 20 per cent chance of spotting the interceptors. The sun, of course, is a great thing to hide in, as anyone looking into it is badly dazzled, and the chances of anyone observing an opponent lurking up-sun are greatly reduced. It should always be plotted on the table. Observation of formations of aircraft is relatively easy up to five miles, and diminishes rapidly beyond this. If, then, we keep it simple and work to the following table, it will come out about right. Up-sun is defined as being within one mile on either side of a line drawn between the target formation and the sun, regardless of range, and with an altitude advantage.

Observation probability table

Less than 5 miles range		5 to 7 miles range		Less than 5 miles range		5 to 7 miles range	
Interceptors				**Raiders**			
Normal	Up-sun	Normal	Up-sun	Normal	Up-sun	Normal	Up-sun
100	05	50	No chance	50	05	25	No chance

Once a unit is observed, it is kept under observation unless it vanishes into the up-sun position when it is lost, and must be diced for again. A detail here, bombers and fighters were almost never on the same wavelength, nor were fighters of different units, although with fighters the distinction is not altogether clear. It would, however, be reasonable to suppose that the aircraft in a *geschwader* in formation would all be on the same wavelength, but those in individual *gruppen* of the same *geschwader*, although in the air at the same time but on different missions, would not. By the same token, British squadrons flying as a wing would share a common frequency, even though on a later trip on the same day they might fortuitously meet in the air and be unable to communicate. What I am saying is that if one force spots something, be careful to consider who it can reasonably warn, barring of course the ground controller, who is told everything and can tell

Large-scale air warfare 135

everyone else, time permitting. Also bear in mind that there was rarely a ground controller on offensive trips; you were on your own.

For the interception phase, attacks can be made from many different directions, and an unobserved attack from the rear is always the most deadly. In many books on air warfare it is noted that the damage was usually done by the same few pilots. I have therefore reproduced this for the game, by postulating three grades of pilots. These are the *experte* who is a fine shot, the average, who gets occasional results, and the bat, who is blind as the proverbial, whose results are largely a matter of sheer luck. Formations of three fighters are comprised of bats, four or five fighters contain one average and three or four bats, six or seven fighters contain one *experte* and five or six bats, eight, nine or ten fighters contain one *experte*, one average, and six, seven or eight bats, 11 or 12 fighters contain one *experte*, two average and all the rest are bats. I prefer the German word *experte* to 'ace', in case you are wondering.

A little clarification is needed before we proceed. The intercept phase, lasting a whole move is not quite as clear-cut as it at first appears, due to the fighter/fighter element. Bombers are always classed as intercepted, fighters are intercepted if they are surprised, but if they spot an attack coming in, a section can be detached to intercept an attack in turn. We therefore have a double intercept, or dogfight situation. If, however, the intercepting formation detaches a section to take on the escorts who in turn have been detached to intercept them, then the undetached section will get through unless there is yet another lot of escorts about to intercept them in turn. Each unit can split once. Any more than this and you will need headache pills. A formation of less than six aircraft may not split at all. A little care is needed here in defining formations, as a *gruppe* would normally be composed of three *staffeln* and, of course, can easily be split into three. The *staffel* is therefore the formation that can only be halved.

Interceptions take place from above and below, exactly level counting as above, and from head-on, beam, quarter and astern. These areas are exactly defined in the accompanying diagram.

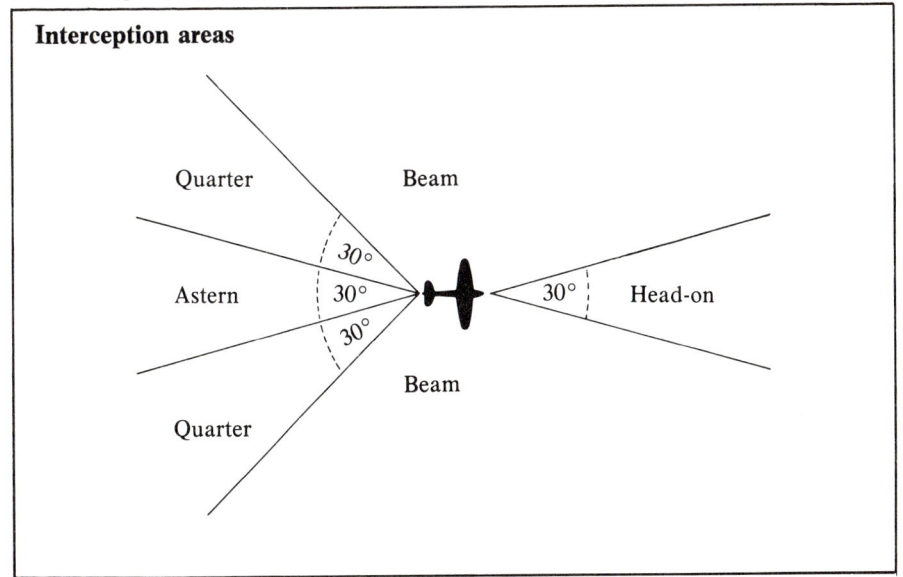

Interception areas

They are also classified as surprised or unsurprised. In a surprise interception, no return fire is possible by the defenders. The percentage dice are thrown once for each *pair* of aircraft attacking; naturally the best shots will be included or, if the attacking unit was flying in vics of three, one dice throw for every *three* aircraft attacking. This is to penalise outmoded tactics. The following table shows the percentage probability of target aircraft being hit:

Interception table

	Above			Below		
	Experte	Avge	Bat	Experte	Avge	Bat
Head-on	20	10	00	16	08	00
Beam	10	05	00	08	04	00
Quarter	30	15	03	24	12	02
Stern	50	25	05	40	20	04

Hits having been scored, the dice are thrown for each hit to ascertain the result, which is done from the following table. The fire factors are those from the aircraft performance data for the skirmish game.

Hit result table

Number of engines in hit aircraft

Fire factor	Result	1	2	3	4
1- 5	Shot down	00-30	00-23	00-15	00-08
	Severe damage	31-53	24-49	16-43	09-39
	Medium damage	54-76	50-74	44-71	40-69
	Light damage	77-99	75-99	72-99	70-99
6-10	Shot down	00-50	00-38	00-25	00-13
	Severe damage	51-67	39-59	26-50	14-42
	Medium damage	68-83	60-79	51-75	43-71
	Light damage	84-99	80-99	76-99	72-99
11-20	Shot down	00-60	00-45	00-30	00-15
	Severe damage	61-73	46-63	31-53	16-43
	Medium damage	74-86	64-81	54-76	44-71
	Light damage	87-99	82-99	77-99	72-99
20-30	Shot down	00-70	00-53	00-35	00-18
	Severe damage	71-80	54-69	36-57	19-45
	Medium damage	81-90	70-84	57-78	46-72
	Light damage	91-99	85-99	79-99	73-99
30 plus	Shot down	00-80	00-60	00-40	00-20
	Severe damage	81-87	61-73	41-60	21-47
	Medium damage	88-93	74-86	61-80	48-73
	Light damage	94-99	87-99	81-99	74-99

Severe damage means that the aircraft sustaining it cannot return to base if the distance exceeds 75 miles. If it does succeed in returning, it will be unservicable for the following ten days. If it crash-lands in friendly territory it is written off. Medium damage means loss of servicability for seven days. Light damage means unserviceable for two days. Aircraft receiving severe or medium damage immediately break off and go home. Rough and ready I know, but it seems to work.

Having carried out a successful interception, the intercepting formation has two choices. It can dive away and escape provided that either its aircraft are ten miles per hour faster than its opponents', or that the interception was made in a dive from at least 2,000 feet higher than the enemy. Or the interceptors can stay and fight. If, in the intercept, a unit has more than 20 per cent of its aircraft lost, or suffering severe or medium damage, it loses a certain amount of cohesion, and fighting effect. This gives its opponents an extra half as much again on their probability of scoring hits in the dogfight phase.

Two things count in fighter versus fighter combat in the dogfight phase: turning ability and speed. Turning ability, as we have seen earlier, is largely dependent on wing loading and can be assessed accordingly. Speed is assessed on the maxium level speed in the aircraft data tables. What we have to do is assess fighter versus fighter, deducting the lowest score from the highest, and taking the difference to assess the aircraft in the combat situation. Score one point for each 25 mph or part that one aircraft is faster than the other, and one point for each 5 lb or part of wing loading that one is less than the other. As an example, let us compare the Hurricane I with the Bf 109E. Comparative level speeds are 354 mph for the '109 and 316 mph for the Hurricane = 38 mph difference = two points for the 109. Comparative wing loadings are 32 lb for the '109 and 26 lb for the Hurricane = 6 lb difference = two points for the Hurricane. Therefore in a dogfight situation they are classed as equals and rated at zero. Anyway, on to the dogfight table, for the probability of scoring hits.

Dogfight table

Fighter rating	Experte	Average	Bat	Fighter rating	Experte	Average	Bat
-6	02	01	00	1	23	11	02
-5	04	02	00	2	26	13	03
-4	05	02	00	3	31	15	03
-3	07	03	01	4	41	20	04
-2	10	05	01	5	58	29	06
-1	14	07	01	6	90	45	09
0	20	10	02				

Hits scored are resolved in the same manner as previously described. The only thing not yet covered is bomber guns. These return fire at the fighters as per the interception table at 'bat' standards, but to compensate for the lack of effectiveness of bombers' return fire only one bomber in every five shoots back, albeit with its full fire factor. Hits are resolved as per the hit result table in the normal way.

In the campaign game, unlike the skirmish game, an allowance needs to be made for morale. Morale can be loosely defined as the will to get stuck in. Situations could (and did) arise where a formation, on being confronted with a menacing situation, pulled the plug and went home. The type of situation for which we need to legislate is two-fold. The first is when a unit in action takes unacceptably heavy losses; the second is when an unescorted bomber formation is faced with an enemy fighter force strong enough to inflict unacceptably heavy losses on it. Unacceptably heavy losses can be arbitrarily assessed as at least 20 per cent of the force receiving medium damage or worse. The second case is more problematical, but can be defined well enough for our needs as a probability based on relative strength, with a basic minimum of one intercepting fighter to every two bombers to start the ball rolling. The probability of unescorted bombers turning back when faced with fighters is given in the 'chicken table' overleaf.

Two further points before we leave the subject. A bomber force is considered unescorted when its escorts are already engaged at a minimum ratio of two escorts

Chicken table

Ratio fighters/ bombers	% Probability of bombers turning back	Ratio fighters/ bombers	% Probability of bombers turning back
3:1	100	1:1	70
2:1	90	3:4	60
3:2	80	1:2	50

to each interceptor. If a bomber forces does jettison its bombs and do a bunk, the interceptors will always pursue, fuel permitting.

Well, that's the air-to-air stuff out of the way; it is now time to consider the air-to-ground goodies. As stated earlier, it is sufficient for the bombers to overfly most targets for the attack to be considered successful. So long as, say, 80 per cent of the bombers attack in this way, the target may be assumed to have been well and truly bombed. The exceptions, because they are to do with the air game, are radar stations and airfields. Radar stations are small and difficult targets to knock out, as the Luftwaffe found in 1940, and even when knocked out are often back in service within a few hours. Even when a damaging blow has been struck, the gap can be partially plugged the same day with a small mobile set. For attacks on radar stations, throw the dice for every three dive-bombers, or every ten level bombers. A score of 00 puts them out of action. Having succeeded, roll a six spot die, and add two to the score. This is the time in hours for which the station is off the air.

Airfields are a bigger target, but with all the wide open spaces it is difficult to cause much damage unless a saturation attack is mounted. Bombers flying in formation are unable to make more than minor adjustments to their bombing run for fear of collision, therefore they can really only let fly at what is under their noses. If they were to break formation and pick targets individually, their losses would be prohibitive, especially if enemy fighters came on the scene while the formation was split up. Flak would tend to make them hurry their aim, and if enemy fighters were attacking during the bombing run, their concentration on the target would suffer badly. Finally of course, observation rules apply. If you don't see the target, you don't attack, although you can always come down below the cloud.

As stated, an airfield is a pretty spread-out target. Even such vast areas as the runways rarely occupy more than two per cent of the total area. The best targets are, in order of precedence, aircraft on the ground, hangars, and runways.

Fuel and ammunition are not particularly relevant to our game. They should, in theory anyway, be dispersed, and so well camouflaged as to be indistinguishable from the air, making a hit a matter of pure chance at astronomical odds. Even if a hit were to be obtained, no more than a local temporary shortage is achieved, which is quickly remedied by the ground supply organisation. We can therefore feel justified in omitting this altogether. Parked aircraft are quite a different matter. They may be unserviceable aircraft which will be repaired in a day or so, in which case their destruction will quickly be made good by the supply organisation. They may, hopefully, be the reserve aircraft of a squadron in action somewhere else, which on returning to base with a few aircraft damaged or missing, will not be able to make up the deficiency from stock, and must fly at reduced strength for the rest of the day if their reserves are destroyed. Or they may—Oh joy! Oh rapture!—be a complete operational squadron caught on the ground. To attack grounded aircraft, proceed as follows:

1 Decide what proportion of your bomber force is to attack the grounded aircraft.

2 Turn to the airfield attack table and evaluate the probability of scoring a hit.

Large-scale air warfare

This is done by defining the target, let us say aircraft in the open, the immediate opposition, say airfield flak, and the height at which you are flying, let us assume 12,000 feet. Under the heading 'targets for level bombers', we find a sub-heading of 'immediate opposition'. This is further divided into three columns headed 'nil', 'flak' and 'fighters'. As we are under fire from flak, we look down this column until it crosses the 12,000 feet line in the 'attack height' column and read off the percentage probability of scoring a hit under these circumstances of 04.

3 Roll the dice for each bomber in the formation; scores of 04 or less mean hits.

4 Evaluate each hit as follows: roll the dice for each hit scored. The effect is in accordance with the following table.

Dice scores	Effect
00-50	Destroyed
51-70	Severe damage
71-90	Medium damage
91-99	Light damage

Our next target to consider is hangars, which are fairly large, and not too difficult to hit. The same procedure applies. The benefits of hitting them are that any aircraft inside them can be assumed to be destroyed. Also, if you manage to tip a minimum of 4,000 lb of bombs on to any one hangar, the hangar itself is destroyed, which reduces the airfield repair capability. Working on the assumption that on average there are four hangars per airfield, the destruction of all four means that all repair work ceases for 24 hours, and on resumption, work is at half speed for the following week.

Runways are easy targets to hit, but very difficult to mess up completely. The average airfield contained two runways of a total length of 1,800 yards. Assume that a minimum of 10,000 lb of bombs on a runway render it totally unserviceable, and that repairs take half an hour for 1,000 lb of bombs. With the runway

Airfield attack table

Targets for level bombers

Immediate opposition

Attack height	Aircraft in open Nil	Flak	A/C	Aircraft in blast pens Nil	Flak	A/C	Hangars Nil	Flak	A/C	Runways Nil	Flak	A/C
8,000	20	15	10	06	04	03	30	22	15	60	45	30
9,000	16	12	08	04	03	02	24	17	12	48	36	24
10,000	12	09	06	03	02	01	18	13	09	36	27	18
11,000	09	07	04	02	01	00	14	10	07	28	21	14
12,000	07	05	03	01	00	00	11	08	05	22	16	11
13,000	06	04	03	00	00	00	09	07	04	18	13	09
14,000	05	04	02	00	00	00	07	05	03	14	10	07
15,000	03	02	01	00	00	00	05	04	02	10	07	05
16,000	03	02	01				04	03	02	08	06	04
17,000	02	01	00	No dice (literally)			03	02	01	07	05	03
18,000	02	01	00				03	02	01	06	04	03
19,000	01	00	00				02	01	00	05	04	02
20,000	01	00	00				02	01	00	04	03	02

Dive-bombers (standardised attack height)

	27	20	13	09	07	04	40	30	20	80	60	40

unserviceable, fighters weighing four tons or less can generally put down on the grass anyway.

You will note that I have made no provision for low-level attacks. This is because the levels of accuracy would be likely to be much the same as from 8,000 feet, as the time available to evaluate the target is very short.

Dive-bombers I have assumed will release their bombs from a fairly standard height, which will vary according to the opposition encountered. After the attack they will reform at 7,000 feet to go home.

The final point to cover is ground-to-air fire. This must be on a very arbitrary basis, and I would suggest that when final casualties are calculated at the end of the day, five per cent added to lost and damaged categories will suffice.

Carrier battles were mentioned earlier. The best way of handling these is to do the raids, intercepts, etc, all as described, but to carry out the strike all as the skirmish game. The two essentials for carrier battles which differentiate them from the game described so far are reconnaissance and secret movement. Weather and wind direction are sorted out in the usual manner, then the scouting aircraft are flown off, first having been pre-programmed with courses for a search pattern. At this stage, neither carrier force will be shown on the board. Having flown out a minimum of 50 miles, and preferably more, the scouting aircraft will use a search pattern, which will consist of a clear plastic circle, the size of which will depend in what height band the aircraft is flying. The diameters can be arbitrarily assessed as one mile (centimetre) for each thousand feet of altitude. The search pattern is placed with the scouting aircraft central. When the search pattern overlaps the enemy fleet, a sighting is made, weather permitting. As soon as search patterns are used, the fleets are placed on the board, and moved around at a speed of 20 knots, on a pre-determined course. As soon as a sighting is made, the respective commanders are free to react as the situation warrants. A strike takes half an hour to plan and after this time, aircraft can commence taking off, provided that the carriers have turned into wind.

When a strike is to be launched by an unlocated force, the principle of secret movement is prejudiced, as his opponent is able to see from which direction the strike is coming. Having been located, he will realise that he will shortly be on the receiving end, and can dispose his forces to meet it. We overcome this by an extremely simple device. The carrier strike takes off, forms up, and starts in towards the enemy. On reaching a range of 50 miles from the opposing fleet, which is our arbitrary carrier radar range irrespective of height, the position of the force launching the strike, and the line of flight of the strike, can be altered by 45 degrees either side of the original flight line. The launch force remains at the same distance, but changes position to suit. All this is fine provided that a scout, dodging in and out of cloud, has managed to keep contact with the hostile fleet, but normally at least four aircraft would be scrambled to clobber it. If there is a minimum of five tenths cloud it will stand a 90 per cent chance of evading interception per move. If, however, the scout is driven off or shot down, contact with the enemy fleet will be lost. It will change course and generally make off at high speed. In this event, the strike force fly to the place it was last seen. They will then split up and search, but if split up have no radio contact with each other. The search pattern is then used for each formation as before. Directly the search pattern is used, the target fleet will once more be placed on the board.

The best way of doing the attack is, having got any interceptions out of the way, to revert to the skirmish game anti-shipping principle and use models.

Chapter 10

Campaign details

In the previous chapter we covered the essential outlines of mounting either large raids or a full-scale campaign. As requirements in any particular theatre of war tend to vary, it will be as well to have a closer look at the necessities in a couple of specific areas. I have selected the Battle of Britain as being of the greatest interest to the majority, and Malta 1942 as offering the greatest variety of problems to solve. Both of these have one great advantage; a large dollop of sea prevented land warfare from encroaching on them, making them both a nearly pure air war.

As in all campaigns, the considerations are: 1) bases and deployment; 2) organisation and order of battle; 3) radar; 4) turn-round times; and 5) supply situation—aircraft, pilots and stores.

Let us take the Battle of Britain first, starting with bases and deployment. The defence of Britain against air attack was the responsibility of Royal Air Force Fighter Command, which covered the whole country. This was subdivided into Groups. 10 Group covered the West Country across as far as Oxford. 12 Group covered the Midlands and the northern half of East Anglia. 13 Groups took care of the north of England and Scotland, while 11 Group, which bore the brunt of the fighting took in the whole of south-east England, from the Isle of Wight eastwards, including London, and the southern half of East Anglia. As the map-making instructions in chapter 9 dealt primarily with this Group, let us break it down in more detail.

Each Group was sub-divided into sectors. In 11 Group, these were Tangmere, Kenley, Biggin Hill, Hornchurch, North Weald, Debden and Northolt. Each sector was named after the airfield from which it was controlled. This subdivision was necessary as, for technical reasons, only a limited number of squadrons could be controlled from one place. There were other airfields within each sector, but control was always from the sector station. Had the Luftwaffe known this, and also that the control room was on the station and virtually unprotected, they would have hammered the sector stations into dust as quickly as possible, and until new arrangements had been made, the effectiveness of that sector would have been greatly reduced.

Playing the game with the advantage of hindsight must be compensated. We do this by omitting sector stations and controlling our entire battle from a top-secret HQ. Furthermore we need to reproduce another German intelligence failing. Large attacks were often launched against airfields which did not belong to Fighter Command. Hindsight would, of course, obviate this so we need to re-introduce the possibility. This is done by the defender doing a swap, say for example exchanging the fighter airfield at Gravesend for the Coastal Command airfield at Eastchurch. He can do as many swaps as he likes *before* play commences.

The basic unit of Fighter Command was the squadron. With a nominal strength of 16 aircraft, the normal flying strength was 12, although odd numbers such as 13 and 16 have been recorded. It was occasionally ordered off in flights of six, although this occurred more when flying patrols than for interceptions. For our game, we fly in sixes or twelves, unless reduced by unserviceable aircraft. Occasionally wings of two to five squadrons were used, but this was rare.

The Luftwaffe deployment and organisation was basically different. It comprised three *Luftflotten*, which for all practical purposes were balanced air forces in their own right. A *Luftflotte* comprised up to three *Fliegerkorps* of bombers and a *Jafu*, or fighter division.

The largest cohesive flying unit in the Luftwaffe was the *Geschwader*, which occasionally took to the air *en masse*, although it was more usually flown in its smaller component parts. A *Geschwader* normally comprised a *Stab* of three or four aircraft, and three *Gruppen* of 27 aircraft each, although some *Geschwadern* contained two *Gruppen* and some had four. A *Gruppe* comprised three *Staffeln*, each of nine aircraft. Taking an unserviceability rate of 25 per cent, it seems reasonable to suppose that each *Staffel* would put up six aircraft, thus making the flying strength of a *Gruppe* 18 aircraft, and a *Geschwader* 54 aircraft. As the *Gruppe* was the basic flying unit, we can generally discount the *Stabs* unless the *Geschwader* is flying as one formation. The function of *Geschwadern* can be identified by the abbreviations as follows: **KG** = *Kampfgeschwader* = bomber unit; **JG** = *Jagdgeschwader* = fighter unit; **ZG** = *Zerstörergeschwader* = heavy fighter unit; **StG** = *Sturzkampfgeschwader* = dive-bomber unit; and **LG** = *Lehrgeschwader* = also a bombing unit.

In addition there were a few odd-ball *Gruppen* formed for a special purpose, such as EGr 210—*Erprobungsgruppe* 210—test group 210 concerned with fighter-bomber operations, and KGr 100, *Kampfgruppe* 100, which was a pathfinder unit.

The numbering of German units can be confusing to the uninitiated and deserves a brief mention. All *Geschwadern* and independent *Gruppen* are numbered ordinarily with Arabic numerals, for example EGr 210, or JG 26. The *Staffeln* within the *Geschwader* were also numbered with Arabic numerals, from 1 to 9 or 12. The *Gruppen* however, were in Roman numerals, I, II, III or IV. This means that 3/JG 26 is the third *Staffel* and is part of I/JG 26. By the same token, III/JG 26 would be composed of the seventh, eighth and ninth *Staffeln*.

We next come to the forces available to either side. This naturally fluctuated from time to time and the difficulty is to know exactly when to plump for. I could name a handful of dates as supposedly being the commencement of the battle, while one or two of the opposition have actually denied that there was such a thing as the Battle of Britain. Be that as it may, if I settle for August 13 1940, *Adler Tag* as the Germans called it, and quote the order of battle for both sides, no-one can quibble. *Adler Tag*, or Eagle Day, was the day the Luftwaffe set out to destroy the Royal Air Force.

In the following Order of Battle, Blenheim squadrons have been omitted, also Gladiator squadrons. My sincere apologies to those who flew them at the time.

Battered squadrons can be replaced by fresh squadrons from the rear areas. The exchange takes place on the same day, but neither unit sees action on the day of the exchange (except by accident).

Royal Air Force Order of Battle August 13 1940

10 Group

Sector	Sqn	Aircraft	Airfield
Middle Wallop	152	Spitfire	Warmwell
	238	Hurricane	Middle Wallop
	609	Spitfire	Middle Wallop
Filton	87	Hurricane	Exeter
	213	Hurricane	Exeter
Pembrey	92	Spitfire	Pembrey
St Eval	234	Spitfire	St Eval

11 Group

Sector	Sqn	Aircraft	Airfield
Tangmere	43	Hurricane	Tangmere
	601	Hurricane	Tangmere
	145	Hurricane	Westhampnett
Kenley	64	Spitfire	Kenley
	111	Hurricane	Croydon
	615	Hurricane	Kenley
Biggin Hill	32	Hurricane	Biggin Hill
	501	Hurricane	Gravesend
	610	Spitfire	Biggin Hill
Hornchurch	54	Spitfire	Hornchurch
	65	Spitfire	Hornchurch
	74	Spitfire	Hornchurch
	266	Spitfire	Hornchurch
North Weald	56	Hurricane	North Weald
	151	Hurricane	North Weald
Debden	17	Hurricane	Debden
	85	Hurricane	Debden
Northolt	1	Hurricane	Northolt
	257	Hurricane	Northolt
(RCAF)	1	Hurricane (operational 16/8/40)	Northolt
(Polish)	303	Hurricane (operational 31/8/40)	Northolt

12 Group

Sector	Sqn	Aircraft	Airfield
Coltishall	66	Spitfire	Coltishall
	242	Hurricane	Coltishall
Duxford	19	Spitfire	Duxford
(Czech)	310	Hurricane (operational 18/8/40)	Duxford
Wittering	229	Hurricane	Wittering
Digby	46	Hurricane	Digby
	611	Spitfire	Digby
Kirton in Lindsey	222	Spitfire	Kirton in Lindsey
	264	Defiant	Kirton in Lindsey
Church Fenton	73	Hurricane	Church Fenton
	249	Hurricane	Church Fenton

13 Group

Sector	Sqn	Aircraft	Airfield
Catterick	41	Spitfire	Catterick
Usworth	72	Spitfire	Usworth
	79	Spitfire	Acklington
	607	Hurricane	Usworth
Wick	3	Hurricane	Wick
	232	Hurricane	Sumburgh
	504	Hurricane	Castletown
Turnhouse	141	Defiant	Turnhouse
	245	Hurricane	Aldergrove
	253	Hurricane	Prestwick
	602	Spitfire	Drem
	605	Hurricane	Drem
Dyce	603	Spitfire	Dyce

These were the actual dispositions, there is of course no need to stick by them. Not all the satellite airfields are shown in the Order of Battle. I have, therefore, drawn up a list of these, also including airfields with which swaps could reasonably be made, as previously suggested, although I have confined them to 11 Group sectors and the immediate surroundings.

Middle Wallop—Boscombe Down and Worthy Down; Tangmere—Eastleigh, Lee-on-Solent, Gosport, Thorney Island, Odiham, Farnborough, Ford; Kenley—Redhill, Croydon; Biggin Hill—West Malling, Lympne, Hawkinge; Hornchurch—Rochford, Detling, Eastchurch, Manston; North Weald—Stapleford, Martlesham Heath; Debden—Castle Camps, Wattisham; Northolt—Hendon, Heathrow; and Duxford—Cambridge, Fowlmere.

Battle of Britain Airfields 1940

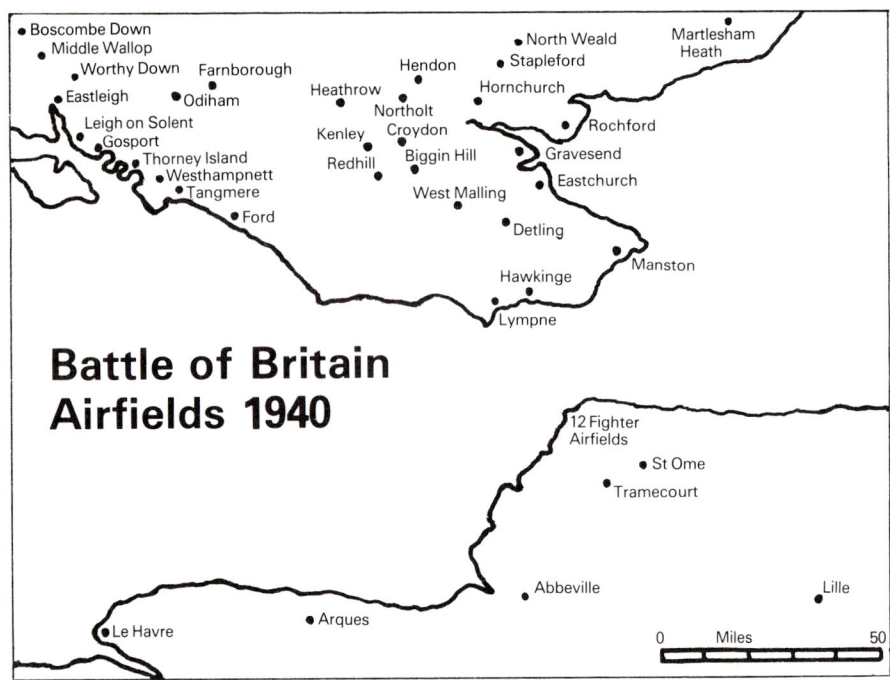

Drawn up against them were three *Luftflotten*. *Luftflotte* 5 consisted of I and III/KG 26, I and III/KG 30, I/ZG 76 and II/JG 77, based in Norway and Denmark. These, if we confine our campaign to south-east England, do not really concern us. *Luftflotten* 2 and 3, based on Brussels and Paris, were the immediate opponents. Their approximate deployment, approximate as I have been unable to ascertain the dates of certain moves, was as follows:

Luftflotte 2

Gruppe	Aircraft	Base
I/JG 3	Bf 109	Samer
II/JG 3	Bf 109	Samer
III/JG 3	Bf 109	Desvres
I/JG 26	Bf 109	Audembert
II/JG 26	Bf 109	Marquise
III/JG 26	Bf 109	Caffiers
I/JG 51	Bf 109	Wissant
II/JG 51	Bf 109	Desvres
III/JG 51	Bf 109	St Omer
I/JG 52	Bf 109	Cocquelles
II/JG 52	Bf 109	Peuplingen
I/LG 2	Bf 109	Calais-Marck
I/JG/54	Bf 109	Guines
II/JG 54	Bf 109	Hermalinghen
III/JG 54	Bf 109	Guines
I/ZG 2	Bf 110	Amiens
II/ZG 2	Bf 110	Guyancourt
I/ZG 26	Bf 110	St Omer
II/ZG 26	Bf 110	St Omer
III/ZG 26	Bf 110	Arques
II/ZG 76	Bf 110	Abbeville

Luftflotte 2 continued

Gruppe	Aircraft	Base
I/KG 1	He 111	Montdidier
II/KG 1	He 111	Montdidier
III/KG 1	Do 17	Rosieres-en-Santerre
I/KG 2	Do 17	Epinoy
II/KG 2	Do 17	Arras
III/KG 2	Do 17	Cambrai
I/KG 3	Do 17	Culet
II/KG 3	Do 17	Antwerp
III/KG 3	Do 17	St Trond
I/KG 4	He 111	Soest
II/KG 4	He 111	Eindhoven
III/KG 4	Ju 88	Amsterdam
I/KG 53	He 111	Lille
II/KG 53	He 111	Lille
III/KG 53	He 111	Lille
I/KG 76	Do 17	Beauvais
II/KG 76	Ju 88	Creil
III/KG 76	Do 17	Cormeilles
EGr 210	Bf 109 and 110	Calais-Marck

Campaign details 145

Luftflotte 2 continued		
Gruppe	**Aircraft**	**Base**
II/LG 2	Bf 110	St Omer
II/StG 1	Ju 87	Calais-Marck
IV/StLG 1	Ju 87	Tramecourt

Luftflotte 3		
Gruppe	**Aircraft**	**Base**
I/JG 2	Bf 109	Beaumont-le-Roger
II/JG 2	Bf 109	Beaumont-le-Roger
III/JG 2	Bf 109	Le Havre
I/JG 27	Bf 109	Plumetot
II/JG 27	Bf 109	Crépon
III/JG 27	Bf 109	Carquebut
I/JG 53	Bf 109	Rennes
II/JG 53	Bf 109	Dinan
III/JG 53	Bf 109	Brest
III/ZG 76	Bf 110	Laval
KGr 100	He 111	Vannes
I/KG 27	He 111	Tours

Luftflotte 3 continued		
Gruppe	**Aircraft**	**Base**
II/KG 27	He 111	Dinard
III/KG 27	He 111	Rennes
I/KG 51	Ju 88	Melun
II/KG 51	Ju 88	Orly
III/KG 51	Ju 88	Etampes
I/KG 54	Ju 88	Evreux
II/KG 54	Ju 88	St André
I/KG 55	He 111	Dreux
II/KG 55	He 111	Chartres
III/KG 55	He 111	Villacoublay
I/LG 1	Ju 88	Orleans
II/LG 1	Ju 88	Orleans
III/LG 1	Ju 88	Chateaudun
IV/LG 1	Ju 88	Orleans
I/StG 1	Ju 87	Angers
III/StG 1	Ju 87	Angers
I/StG 2	Ju 87	St Malo
II/StG 2	Ju 87	Lannion
I/StG 77	Ju 87	Caen
II/StG 77	Ju 87	Caen
III/StG 77	Ju 87	Caen

As you can see, to reproduce the entire Battle of Britain as a campaign would be a considerable undertaking. It is, however, possible to produce a scaled-down version in a reasonable form by pitting *Luftflotte* 2 against 11 Group, and by omitting the Debden sector from 11 Group we are able to keep the entire operating area on the map described in the preceding chapter, with the exception of most of the German bomber airfields. As, however, we lose nothing by letting the bombers appear over the baseline from the right direction, this is acceptable.

RAF squadrons based in 10, 12 and 13 Groups can be used as reinforcements, as can German *gruppen* attached to *Luftflotten* 3 and 5. When a reinforcing unit is brought in, neither the reinforcement unit nor the unit being replaced can be used for that whole day. In the case of redeployment, it would be reasonable to suppose that the units shown in the Order of Battle represent the maximum complement of the airfield at which they are based. It is, of course, undesirable to have too many units based on one airfield, even if it is big enough to hold them, as it takes a clear move for each unit to take off.

Turn-round times are the next matter for discussion. A unit of up to 12 aircraft coming in to refuel needs to spend only two clear moves on the ground before it is ready to go up again. If, however, a unit has been in action, it will not only need to re-arm as well as refuel; it will most likely straggle back in ones and twos, and need five clear moves on the ground before going up again. Should a unit, through lack of fuel, land back on an airfield which is not its base, it will need three moves to refuel, then fly back to its home base to re-arm and re-organise. If, in doing this, it exceeds the base capacity and the home unit is re-fuelling or re-arming, the visitor must wait its turn.

Replacement aircraft and pilots did not cause the Germans many problems. A written-off aircraft would generally be replaced in a couple of days, as would a pilot. Normally there were enough reserve aircrew with the *gruppen* to fill any gaps, although a consistent loss rate of 12 per cent would have been unacceptable. The story was potentially different with the RAF. Although 25 per cent of shot-down pilots survived unhurt and were ready for action on the following day, the pilot replacement situation was always serious. Starting the battle with approximately a 25 per cent surplus, only six pilots per day could be replaced from the Operational Training Units. The aircraft situation was rather better. Starting with a couple of hundred in reserve, the output was about 14 per day, much higher than the Bf 109 output at this time, but the Germans had a bigger reserve of aircraft to start with.

Worthwhile targets, apart from airfields and towns generally, were the Royal Navy at Dover and Portsmouth, and a few aircraft factories, Saunders-Roe at Ryde, Supermarine at Southampton, Folland at Hamble, Short at Rochester, Vickers at Weybridge, Hawker at Kingston and Handley Page at Cricklewood. Total destruction of any of these would have been nasty but not decisive, as fighter production was rapidly being switched to the midlands and north.

Malta was a different kettle of fish and different factors must be considered. First it was under attack by both the Luftwaffe and the Regia Aeronautica. We have described the organisation of the Luftwaffe earlier, so let us fill in on the Italians.

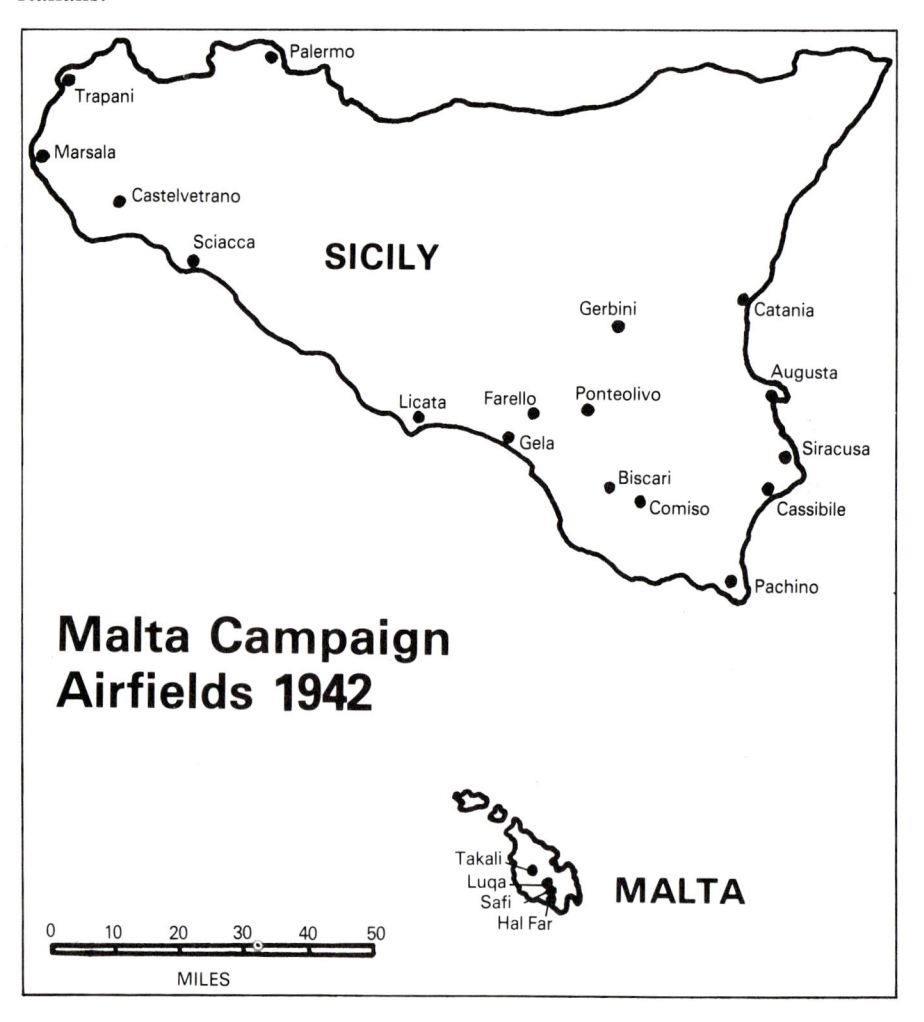

Malta Campaign Airfields 1942

Campaign details

The basic flying unit was the *squadriglia*, which was composed of nine aircraft. Three *squadriglie* made up a *Gruppo*. Thus far, the organisation is identical to the Luftwaffe. However, only two *gruppi* made up a *Stormo*, which is the largest flying unit, and would generally consist of a nominal strength of 54 aircraft, and a typical operational strength of 36. *Gruppi* were often detached to operate independently of a *Stormo*, in which case they were called a *Gruppo Autonomo*. Role suffixes were also used; the most common ones are listed below.

CT = *Caccia Terreste* = fighter; **BT** = *Bombardamento Terreste* = bomber; **BaT** = *Bombardamento Tuffatori* = dive-bomber; **A** = *Assalto* = ground attack; and **AS** = *Aerosiluranti* = torpedo bomber.

Malta was always a bit of a rum 'un for the Axis. No sooner did they assemble a really devastating line-up than it was scattered to the four winds due to the need to reinforce other fronts. Thus the Order of Battle was constantly changing regardless of the needs of the moment. If we then use the actual units and movements, the Axis have a built-in handicap. The following is as nearly accurate a chronological list as I have been able to establish after exhaustive (and exhausting) enquiries. It commences in January 1942. 1940 and 1941 are even more confused so regretfully I am not attempting to give them. The Luftwaffe units listed on January 1 1942 arrived in Sicily in mid-December 1941, mainly from Russia. The Axis forces at the start of 1942 were as follows:

Four *gruppen* of Bf 109Fs: II/JG 3, I, II and III/JG 53. III/JG 53 left for Africa immediately; one night fighter *gruppe*, I/NJG 2, equipped with Ju 88cs based at Catania; one heavy fighter *gruppe*, II/ZG 26, with Bf 110cs; four Stuka *gruppen*, Egr/StG 1; I, II and III/StG 3, with Ju 87Bs. I/StG 3 left for Africa immediately; five bomber *gruppen*, equipped with Ju 88As. They were I/KG 54 at Gerbini, II and III/KG 77 at Comiso, and KGr 606 and KGr 806, both at Catania.

The Regia Aeronautica contribution was small, consisting of 37° *Stormo*, made up of 40 and 55 *Gruppi* with Breda BR 20s for night bombing. 55 *Gruppo* was at this time starting to re-equip with the Cant Z 1007 *bis*.

No real changes took place until April, III/JG 53 returning from the desert on the 1st. Also on this date II/StG 3 started to re-equip with Ju 87Ds. On the 14th 4° *Stormo*, with Macchi MC 202s, arrived back from the desert. Late May saw wholesale changes as the Luftwaffe departed for pastures new. Off to the desert went III/JG 53 (yes, again) together with I/NJG 2, III/ZG 26, II/StG 3. I/JG 53, II/JG 3, and II and III/KG 77 left for the Russian front, and the exodus was completed by the departure of I/KG 54 for Greece. As these units left, their places were taken by Italian units. They were 2 *Gruppo* CT with Reggiane 2001s, 155 *Gruppo* CT with Macchi 202s, 102 *Gruppo* BaT with Ju 87Bs, 4, 33 and 88 *Gruppi* BT, equipped respectively with SM 84s, Cant Z 1007s and BR 20s. June saw the mass exit continue with the departure of 4° *Stormo* for Africa, also 40 and 55 *Gruppi*. In mid-July, the entire Italian contingent which had arrived in May went elsewhere, leaving only four German units holding the fort. The trend reversed in August with the arrival of I/JG 77 from Russia, and during September a fair-sized build-up took place, comprising no less than eight *gruppen* of Ju 88s, I and III/KG 26, III/KG 30, I, III and III/KG 54, II/KG 60, and I/KG 77, three *gruppi* of Macchi 202s, one *gruppo* of Reggiane 2001s, one *gruppo* of Ju 87Bs, and three *gruppi* of Cant Z 1007s. The Axis replacement aircraft situation was none too good. If we work on the assumption that each unit arrives at full strength, but due to the poor supply situation serviceability is two thirds for the Luftwaffe and only half for the Regia Aeronautica, it should work out about right.

British strength varied widely during the year, but its offensive capacity was always limited by the number of aircraft that could survive. Bomber units tended to be withdrawn to Egypt when the going got hot, returning when the pressure slackened. Fighters could in theory be assessed in terms of squadrons, but there is not a lot of point in doing this when five fighter squadrons can muster a mere half dozen serviceable aircraft between them. The strength of the Royal Air Force on Malta must thus be reckoned in terms of serviceable aircraft rather than units. There was always a surplus of pilots, and often a pilot would have to wait ten days or more between flights.

At the start of the year, British strength on the island comprised 110 Hurricanes (81 serviceable), 36 Blenheims, 40 Wellingtons, 25 Swordfish and Albacores, and 16 Marylands for reconnaissance, based on three airfields: Takali, Luqa and Hal Far. Luqa and Hal Far were connected with a taxi-way, which ran through Safi Strip; an emergency landing ground for fighters only. Later on, Safi was turned into a decoy and tastefully decorated with written-off aircraft all carefully dispersed to appear to the morning recce aircraft to be fully operational. This needs to be reflected in the game by making Safi a full scale airfield and using one of the others as a decoy. A further airstrip had been started at Krendi, but this never reached the stage of being useable; its only function was to act as a decoy for night raids. There was also a seaplane base at Kalafrana, but this need not really concern us.

Aircraft replacements for the RAF are easy to sort out. Blenheims in transit to the Middle East were purloined by the Commander-in-Chief, Air Vice Marshal Hugh Pugh Lloyd, who earned himself the soubriquet of 'Sticky Fingers' by his activities, although I don't think he ever managed more than one or two a week. Generally though, bomber squadrons flew in and out from Egypt as the situation permitted. The arrival of a bomber squadron on the island was a signal for the Luftwaffe to do its conkers, frequently to such an extent that a section of three dive-bombers was laid on to obliterate each British bomber. When only fighters were based on Malta, the reaction was never as strong. This was, of course, because bombers spelt danger to the convoys supplying Rommel, whereas fighters did not. Bombers, then, can be flown in when the fighters are strong enough to protect them. When the fighter strength falls, the bombers either evacuate or are destroyed on the ground.

Fighter reinforcements were made by aircraft carrier, Spitfires with long-range tanks being flown in from 600-odd miles away. Fighters, which in 1942 were all Spitfires Vcs, came in as follows: 15 on March 7, nine on the 21st and seven on the 29th; 46 on April 20 and 60 on May 9, these large numbers being made possible by the loan of the USS *Wasp*; 17 more arrived on May 18, followed by a further 27 on June 3, and 32 on the 9th. July deliveries were 31 on the 16th and 28 on the 21st. Three further shipments were made, 37 and 29 on August 11 and 17, and 29 on October 24. After this, extra long-range tanks were used to fly direct from Gibraltar. You will have noticed that a regular flow came in during June and July, when the Axis forces were at their lowest ebb. This enabled strength to be built up to such a pitch that units could be sent off to the desert.

Serviceability was always a problem, and cannibalisation took place on a vast scale. Generally, however, serviceability was on a par with the Luftwaffe. During 1942, there were no hangars left on Malta, so all except the most awkward jobs were done in the blast pens on the airfield. The blast pens themselves had been brought to a fine art, a high proportion of them being made out of empty

Campaign details

four-gallon petrol cans. By April, there were no less than 358 of them; over 200 for fighters, 150 for various sizes of bomber, and two dozen to house the steamrollers and bowsers.

Some indication of what a vast undertaking this was can be given by a look at what was involved in constructing a pen for one Wellington. The pen was three-sided, measuring 90 feet by 90 feet by 14 feet high. It contained *60,000* petrol cans, which were filled with sand and stones weighing over 3,000 tons. The base was 14 cans wide, tapering to two cans at the top. They were wired into position, and took 200 men three weeks to build.

The number of pens to house rollers and bowers may seem excessive, especially as I don't think that one solitary bowser had survived into 1942. But the rollers, used for repairing the runways, were absolutely vital. The average bomb crater took five tons of stone to fill it, and compaction had to be sufficiently good to take a 12-ton Wimpey doing a heavy landing. By 1942, the loss of just one steamroller would have been a major tragedy. For nearly three long years they trundled on, surviving countless near-misses, never faltering even in the darkest hours, answering the massed air armadas of Hitler and Mussolini with a defiant, even jaunty plume of smoke. They had personalities, and having them, were given names, long-lost in the mists of time. I like to think of them as F----, H--- and C------.

A few final details. Aircraft fuel and ammunition was supplied *in extremis* by submarine. This, coupled with the supply ships which did get through, was sufficient to keep the air defences going. Not so the anti-aircraft guns which, except for the guns covering the airfields, were rationed to 15 rounds per day for long periods. It would be reasonable to assume a shortage at all times with the exception of the fortnight immediately following the arrival of a convoy. Ships in harbour were liable to be attacked, but smoke pots were set off on the approach of a raid, which gave effective concealment, although it did nothing for bronchial dockers. Supplies could, when properly organised, be unloaded at 2,500 tons per day. (6,000 tons was the average load.) Turn-round for reinforcement fighters, after a bad hiccup on April 20, was very quick, taking between four and seven minutes per aircraft. From touch-down, each aircraft was marshalled into a pre-designated pen where it was refuelled, checked, re-armed if necessary; the pilot was replaced by a Malta 'old hand' and away it went.

The cause of all the fuss was the Axis supply convoys to North Africa. In the

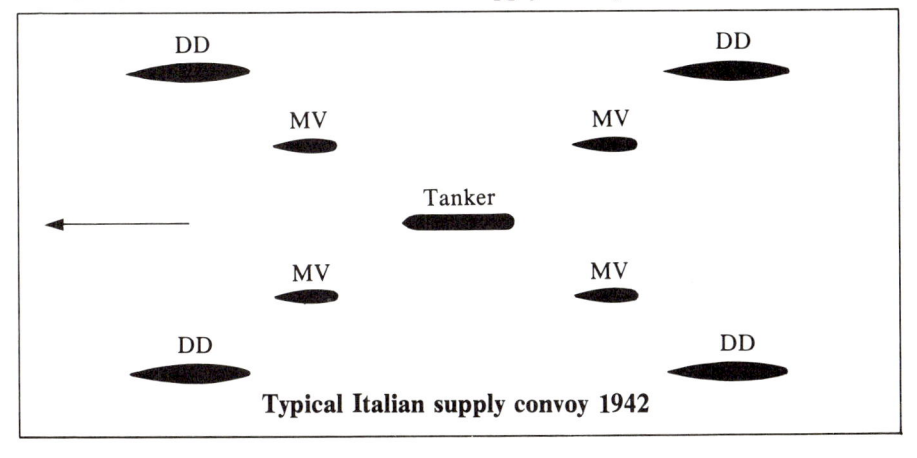

Typical Italian supply convoy 1942

months preceding 1942, losses had been mounting to both air and sea attack. Based on Malta at this time was Force K, consisting of the cruisers *Aurora* and *Penelope*, with two destroyers; also the 10th Submarine Flotilla. During autumn 1941, losses reached catastrophic heights, culminating in 58 per cent during November. The reaction to this was the Luftwaffe build-up on Sicily shown in the Order of Battle; in the face of which our naval forces were compelled to withdraw. The first six months of 1942 saw the trend reversed, with the shipping losses falling to an acceptable six per cent. However, the Axis goofed off elsewhere, Malta recovered its strength, and losses for the second half of the year rocketed to nearly 36 per cent, which arguably had a decisive influence on Alamein, fought in late October of that year, and the subsequent retreat of the Afrika Korps. If you wish to introduce the convoy attacks into your game, you will need to expand your map area considerably, but I would certainly advise you to try it. The average Axis supply convoy consisted of about four ships plus a tanker, with three or four destroyers as escort, deployed as the accompanying diagram. Where within air cover range, they were escorted, serviceability permitting, the air cover usually consisting of four or six Ju 88s or Macchi MC 202s. The convoy routes were Naples/Palermo/Pantellaria/Tripoli, Naples/Reggio Calabria/Benghazi, or Taranto/Benghazi. There were usually about three convoys around at any one time, and tankers were always the prime target. A story exists, possibly apocryphal, of a Swordfish, intercepting a convoy at night, flying slowly up and down whilst the observer shone his torch looking for the tanker!

Chapter 11

Simulating night fighting

The function of the night fighter was pithily described in the early days of the war as like being in the Albert Hall on a dark night with the lights out, looking for someone who isn't there. The bomber wasn't much better off. Granted his target was large and static, but he still couldn't find it unless conditions were exceptionally clear or he was feeling lucky that night. Night flying was an expedient, to be resorted to when daylight operations were likely to produce unacceptably high losses in relation to the probable results. The difficulty was, of course, finding your way about, particularly as a blackout was in operation over Europe. Dead reckoning produced an average error of about seven per cent, and astro-navigation, while quite suitable for ships, was not so good in a relatively fast-moving aircraft as, by the time you had calculated the position on sighting, the aircraft had moved many miles further on. The problems of night warfare then were twofold. The bomber couldn't find the target. The fighter couldn't find the bomber.

The Germans were first in the field to produce aids for night bombing, and had in fact done so even before the outbreak of war. Their *X-Gerät* system, based on the bad weather approach aid developed in the early 1930s, consisted of a radio beam, down which the aircraft flew, with three further beams crossing it. Its maximum range was 180 miles, and the average bombing error was a mere hundred yards provided that the flying was accurate.

Just before the outbreak of war, the Germans developed yet another blind bombing system called *Knickebein*. This was much simpler than *X-Gerät*, using one approach and one cross beam. It did not, however, need special equipment or training, although it lacked the accuracy of the latter. Fortunately for the British, the Germans used these systems prematurely in operational trials; they were discovered and mucked about.

Yet another system, *Y-Gerät*, was introduced in November 1940. This, unlike the other two, was a single beam system. It was discovered almost immediately, jamming was instituted in January 1941, the jamming success rate being in excess of 80 per cent by March.

Whilst the RAF boffins had been enjoying a high rate of success with their countermeasures, their airborne comrades had been enduring a most frustrating time. Gradually, however, an airborne radar set was made to work. The fighter was controlled from a ground radar station and directed towards the bomber. The ground station had a maximum range of about 60 miles and could give a fair indication of height. Friendly aircraft could be positively identified by using their IFF (Identification Friend or Foe) transmitter, which increased the size of the trace in the radar tube. The airborne set, the AI Mark IV, had a two-scope display which showed the target as a blip running along a line, the distance along the line being indicative of range. The operating range was governed by the altitude of the aircraft as the ground returned radar echoes which were picked up by the set and shown on the scope. An aircraft flying at 5,000 feet would have a range of about a mile, whereas at 16,000 feet the range would be three miles.

Contrary to popular belief at the time, the Germans were well advanced in radar. By the summer of 1940 they had two types of ground radar operational: the *Wurzburg*, which had a short range (only 25 miles) but was accurate enough to allow predicted AA fire; and the *Freya*, with a maximum range of 75 miles, and a 360 degree arc. *Freya*, however, could not give any indication of altitude. The RAF bombing raids over the first two years were, I'm afraid, pretty pathetic, only a small proportion of bombs landing within five miles of the target, and many aircraft were unable to find something which even looked like the target.

By the winter of 1941, the Luftwaffe night fighter force was really beginning to get organised. A belt of *Himmelbett* stations was set up reaching from Denmark to Switzerland. Each station consisted of a *Freya* early warning radar, and two *Giant Wurzburgs*. The *Giant Wurzburg* was an improved version of the earlier set; its range was now about 40 miles. One fighter could be controlled from each *Himmelbett*. A bomber would be picked up at long distance by the *Freya*, and passed on to a *Wurzburg* when it came within range. The other *Wurzburg* was used to track the fighter, and gradually the two plots were brought together until the German fighter gained a visual contact.

In the spring of 1942 the defences were further improved by two better early warning devices, *Mammut* and *Wassermann*. *Mammut* had a 200-mile range and a 100 degree arc, but still gave no altitude information, while *Wassermann* had a 360 degree arc, a range of up to 150 miles, and was very accurate on altitude. With these came the *Lichtenstein* airborne radar, first fitted in the Ju 88C. Its maximum range was about two miles, and its display was similar to that of the AI Mk IV.

Meanwhile, the RAF was preparing problems for the German defences. In March 1942 was introduced the GEE navigational system. This consisted of three transmitter stations in England which laid a radar grid across Europe; the accuracy was of the order of 1½ per cent, or three miles in 200, which was a great improvement on previous standards. Then in May 1942 took place the thousand bomber raid on Cologne, in which the aircraft were scheduled to cross the target in 2½ hours. This concentration of force swamped the *Himmelbett* system, as, for every attempted interception, several dozen bombers would get through.

The operational life of GEE as a target-finding aid was short. The German counter-measures organisation started jamming in August 1942, and GEE was rendered impotent fairly quickly. However, during the six months of its life, with the introduction of the bomber stream, the Royal Air Force had started to hit its targets and, what is more, hit hard. Bomber losses fell appreciably, but in war nothing stands still. This was the period when *Lichtenstein* was introduced on a large scale and, by autumn, losses were increasing. Towards the end of 1942, a new device called 'Mandrel' entered service. This caused interference on the *Freya* screens and reduced the effectiveness of the German early warning system.

At the same time, jamming of the German night fighters' communications began. This was done with 'Tinsel', a microphone in the engine bay of the bomber, transmitting a healthy roaring noise on the frequency used by the *Himmelbett* controllers. As the night fighters were almost completely dependent on instructions from the ground during the early phase of an interception, this was quite effective.

At the turn of the year, two new bombing aids entered service, 'Oboe' in December 1942, and 'H₂S' in January 1943. 'Oboe' was a precision bombing aid, similar in principle to *X-Gerät*. Its range was about 250 miles, and it could only handle one aircraft at a time. Its accuracy was excellent. 'H₂S', which was just a code-name, and nothing to do with hydrogen sulphide, was an advanced radar set which gave a sort of television picture of the countryside over which it was flying.

'Oboe' was used for precision attacks, and also by the Pathfinders marking for the main force. Carried in Pathfinder Mosquitoes, it was a year before the Germans found a method of jamming it. Directly jamming was instituted, the system switched to centimetric radar, which at that time could not be mucked about, although the original signals were transmitted to give the German countermeasures organisation something to play with, a fact that the Germans did not discover until July 1944!

'H₂S' was a centimetric radar, and for the duration of the war was beyond the German jamming capability. They could, however, home on its transmissions from quite a distance, and *Naxos*, a radar detector, was quickly developed to do this, entering service late in the year, thus making interceptions easier.

Two other devices were fitted to bombers in the spring of 1943; 'Monica' and 'Boozer'. 'Monica' was a tail warning radar. It could not, however, differentiate between friendly and hostile aircraft and in practice gave so many false alarms that it was not much use. 'Boozer' was a receiver tuned to the emanations of the *Wurzburg* and *Lichtenstein* radars. With the ether full of radiations from different sources, its value would appear to be questionable. Not so 'Monica', which was an absolute menace. The Germans got hold of one and in double-quick time built *Flensburg*, a homing device. The existence of *Flensburg* was not discovered until mid-1944, when all 'Monicas' were scrapped as quickly as possible.

So far, we have concentrated on bombing gadgets, but the RAF had also made great progress in fighter radar. The AI Mark VIII was in service, giving an egg-shaped area of cover ahead of the aircraft, with a maximum range of six miles. This was about to be followed by the AI Mark X with a semi-circular area of cover, and a nine-mile range. Also coming into service was 'Serrate', a passive homer for detecting *Lichtenstein* emissions.

The next weapon to be introduced to the night war was 'Window'. Strips of aluminium foil cut to half the wavelength of the German radars, they produced an echo on the German screens like a heavy bomber. 'Window' was introduced in July 1943, during a raid on Hamburg. For the first time, the *Wurzburgs* and *Lichtensteins* were swamped with a mass of echoes. But the Luftwaffe did not stay on the floor long. They were quickly back on their feet and hitting hard. The *Himmelbett* system was rapidly abandoned. In its place came two methods which, while they were only expedients, brought results.

The first of these was *Wilde Sau*, which comprised up to a hundred single-engined day fighters. Navigating by a grid of radio beacons established over the length and breadth of Germany, these intercepted actually over the target, where the bombers could be seen silhouetted against the light of the flames below. The flak was restricted to a certain height, and the fighters operated above this.

Zahme Sau was the other method. Night fighters were ordered to orbit radio beacons in the projected track of the bombers. Ground control was minimal, being little more than a running commentary on the movements of the head of the bomber stream, with the idea of inserting the night fighters into it. Once there, they

For the first time we use 1:300 scale models instead of the 1:144 half models seen earlier This means that, while still playing in the vertical plane, the aircraft appear in plan view. Don't be put off by this; personally I prefer 1:300 scale. This sequence has been put together to illustrate the two basic fighting formations, the pair and the four, sometimes called the 'finger four'. **1** Here we have four Hurricanes flying a 'finger four'. It is vertical rather than horizontal, but it still works. Behind them and closing fast is a *rotte*, or pair, of Bf 109s.

2 The *rottenfuhrer* goes for the top Hurricane, which breaks upwards, radioing for help. The leading '109 closes on his tail, but the rest of the Hurricanes also break upwards. The British No 2 automatically covers his leader's tail, and the German No 2 does a similar job.

Simulating night fighting 155

3 The Hurricane leader pulls right round over the top, out of harm's way. The '109 leader, with the second Hurricane on his tail, breaks hard down, while the German No 2 slots in behind them, thus exposing his rear to the third Hurricane. 4 The second Hurricane, feeling distinctly nervous about the second '109 on his tail, also breaks up and away, leaving the argument to be settled by the third Hurricane. By this time, the Hurricane leader has rejoined.

were on their own, although the increasing number of passive detectors carried made their task easier, as did the clear summer nights of 1943.

From October 1943, the *Lichtenstein SN2* radar was introduced. Working on a different wavelength to its predecessor, it was unaffected by the 'Window' currently in use. Its range was about four miles, and once again the night fighters were back in business. By this time though, feint attacks and misleading courses were Bomber Command's stock in trade. Sometimes the night fighter controllers could be completely misled; at other times they were not.

By 1944 jamming of the German ground-to-air communications had become so effective that they were forced to desperate measures. One of their broadcasting stations played music to guide the night fighters; jazz for Berlin, waltzes for Munich, etc. Even this expedient was soon recognised and jammed by a special high-powered transmitter.

In the summer of 1944, a new German ground radar appeared, which was designed to beat jamming by operating on any one of four different frequencies. This was *Jagdschloss* which had a range of 90 miles. By this time, the invasion of Normandy had taken place, and a large gap had been knocked in the German early warning system. From September onwards, raids were frequently routed over northern France to take full advantage of this. Concurrently, the Allied bombing offensive against oil targets began to take effect, in the form of an acute fuel shortage, which even grounded operational units which would otherwise have been flying. To add insult to injury, 100 Group commenced operations at about this time, their function being to insert the proverbial spanner into Luftwaffe communications and radar, and to provide an intruder fighter protective screen for the bombers, composed of Mosquitoes fitted with AI Mark X, a 'Serrate' modified to pick up *SN2* emissions, and 'Perfectos', which was surely the most unsporting trick of the war. All you had to do was press a button, and receive an immediate message which said 'I am a Luftwaffe aircraft. I am three miles away on bearing 270 degrees. Please come and shoot me down.' Well nearly, anyway! 'Perfectos' triggered the IFF of any German aircraft in the vicinity. Fortunately for them, the Germans twigged on what was happening very quickly, and used the simplest possible countermeasure. They switched off their IFF.

Other feats of electronic wizardry were introduced, among them 'Jostle', the biggest and most powerful communications jammer of them all, and 'Piperack' which played havoc with the *SN2* radar. The H_2 S detector, *Naxos*, was found to be useful for detecting AI Mark X emissions, but from the autumn of 1944 until the end of the war there were so many Mk X-equipped Mossies around, that all *Naxos* could do was make the night fighter pilots nervous, as the warnings hardly ceased the whole time they were airborne. And so the defence just petered out. The achievements of the German night fighter arm had been beyond praise. But outnumbered and finally outfought, they lost.

The reproduction of night air warfare contains two main difficulties. The first is of giving the player only the information to which he would have had access in reality. The second is that night interceptions were notoriously difficult to achieve and, if reproduced with any degree of accuracy, the level of boredom would become unacceptable. So once again we need lots of compromise.

The first requirement is to set a scenario. This, as you may have gathered from the preceding section, can be difficult. The basic decision is at what level of complexity you wish to play and whether or not you wish to use an umpire to look after the hidden movement. The snag of using an umpire, provided you can get

Simulating night fighting 157

one, is that you have to have absolute confidence in his judgement. If you have such a paragon handy, you can go in for an extremely complex game, but otherwise you are forced to keep it simple. The more simple the game, the faster moving it is, which often means that the level of excitement is high. Personally, I would always plump for the simple game, provided that reality does not suffer too badly from over-simplification.

The night sky is a large place. Looking for odd bombers widely scattered is not much fun, therefore I decided to settle for using the RAF bomber stream over Germany as a target, ignoring their bombing function entirely. To the bomber stream, we add some German night fighters, referred to from now as interceptors, whose purpose is to shoot down bombers. We then chuck in a few British night fighters, from now on called intruders, and we have a fighter versus fighter game.

Our next step is to decide on a playing surface. As pilots flying at night were very dependent on instruments, they tended not to indulge in wild manoeuvres. Their flying was therefore much more restrained than that of their daylight counterparts. We are thus not dependent on marginal aircraft performance differences in the same manner as in the daylight game, with the result that we can use a hexagonal grid surface of the type marketed by Simulation Publications (UK) Ltd, although it does need to be the type in which the hexes are numbered, preferably short grain. Two of these are needed, one for each player. A simple screen of books can be put up between the two boards so that neither player is able to see the actual dispositions of the other. This is the secret movement taken care of. For aircraft, we can again raid SP products; this time for the counters from their game 'Spitfire', which are more than adequate for our needs.

We next need to decide a suitable game scale. The average speed of the bomber stream was in the 180-220 mph region. The night fighters were capable of speeds of 360-400 mph, although when actually closing to intercept, their speed would reduce to little above that of the bombers, because of the danger of overshooting or colliding in the darkness. The bomber stream flew as a general rule in the height band 17-22,000 feet. As changes of altitude on the part of the night fighters tended to be made in a fairly gentle way, with no dramatically steep climbs or dives, we can reasonably ignore the height differential of individual bombers in the stream and play the game entirely in two dimensions. The final requirement is to keep the game within the confines of the playing area. In order to do this, the bomber movement allowance must be two hexes. This, at a speed of 180 mph gives us a hex scale of 880 yards and a move time of 20 seconds. As bomber turning points were few and far between, we can stipulate that the bombers must fly a straight course across the board. They are, after all, only targets. A suitable number of bombers would be 18, of which only 12 are on table at commencement. The bomber stream needs to be a minimum of five miles, or ten hexes, wide and at the start of the game should extend no further from the edge than 15 hexes. They should be randomly distributed; close formation flying is not on, and I would suggest that the area of a radar screen, (with which we shall deal in a minute) should not contain more than five bombers. The bombers which start off-table are allowed to enter at the rate of one per move, or up to three can be saved to enter together. For the fighters, a reasonable balance is given by having four interceptors and three intruders. Starting with an extra interceptor makes the intruders concentrate on their primary task of protecting the bombers rather than hunting down the interceptors.

Fighter movement is a bit more complex. These can move at variable speeds of two, three or four hexes per move. Acceleration/deceleration takes place at the rate

Use of radar screen

Radar screen

Bomber 'echoes'
1811
1709 ← 'Flash' hex No
1606

Radar screen

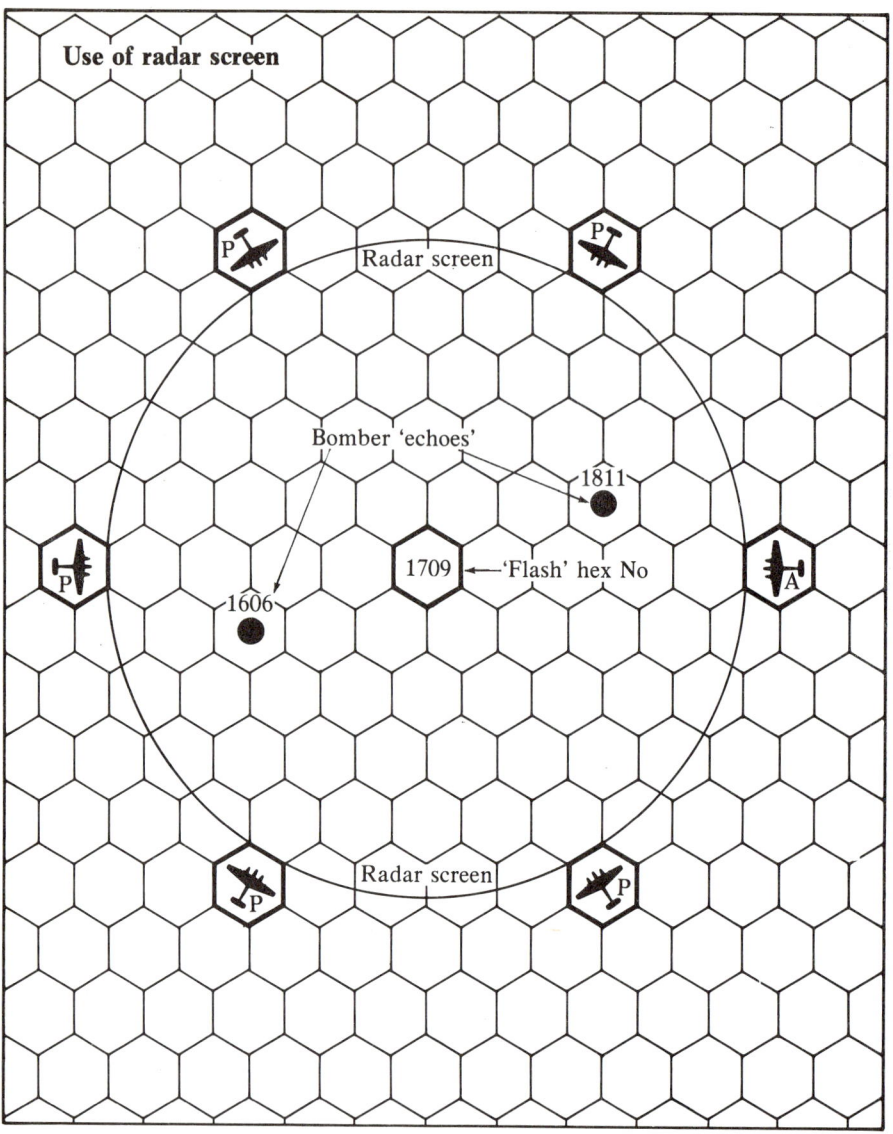

P = Possible night fighter position
A = Actual night fighter position

All hexes are normally numbered but these have been omitted for clarity

of one hex per move, ie, a fighter travelling at four hexes per move can decelerate down to three in the next move, then down to two in the following move. It will equally take two moves to accelerate back up to four hexes per move. Fighters can also manoeuvre. They can turn one hex side for every hex moved, the turn coming at the end of the movement. Thus a fighter travelling at four hexes per move and wishing to reverse course, goes one hex; turn, one hex; turn, one hex; turn, then finishes the move with one hex straight. It may never turn a hex side before any movement takes place, nor more than one hex side for each hex of movement.

Having thus covered movement, we now need to legislate for locating unseen

Simulating night fighting

aircraft by electronic means. If we base our scenarios in the second half of 1944, we can have the interceptors equipped with *Lichtenstein SN2* and *Naxos*, and the intruders with AI Mark X and 'Serrate'. Neither type of radar is particularly affected by 'Window', and the passive detectors can give an indication of the area in which each other's radar is being used. We overcome the ground/air communications jamming by assuming that the interceptors have already been directed into the vicinity of the bomber stream; all they have to do is to cast around and find it. At this juncture, I would strongly recommend initial placement of the intruders within five hexes of the edge of the board.

Anyway, now for our radar set. This consists of a simple circle of clear perspex of a diameter to cover nine hexes, with a central hex cut out. When using the radar, place the circle touching the hex in which the 'flashing' aircraft is situated, then call out the number of the hex under the central cut-out. Your opponent must then place his 'radar' so that the central cut-out is over the hex number called out, and then tell you the hex number of any of his aircraft which are in hexes covered wholly or partially by the radar circle. It is as well to have a large bag of ludo type counters available, (available at any good toy shop) to plot enemy positions. In using the radar, you are not giving away your exact position; only your general locality, which would be indicated by the passive detectors *Naxos* or 'Serrate' anyway. In the illustration titled 'use of radar screen' you can see what I mean. The 'flashing' aircraft (A) is in hex 1714. The 'flash' hex number called is 1709. From this, the enemy can deduce that you could be in hex 1714, but you could equally well be in hexes 1211, 1206, 1704, 2206, or 2211. It's up to him to guess correctly. Also shown in the illustration are two bombers, in 1606 and 1811. You know now where they are, but you don't yet know in which direction they are going. Or are they intruders? You have got to find out. You will only do this with a further 'flash' on your next move, and unless you are very cunning, this will definitely give away your position to an alert opponent. You do not, however, have to use your radar in every move, although it is compulsory during the move in which you make an interception.

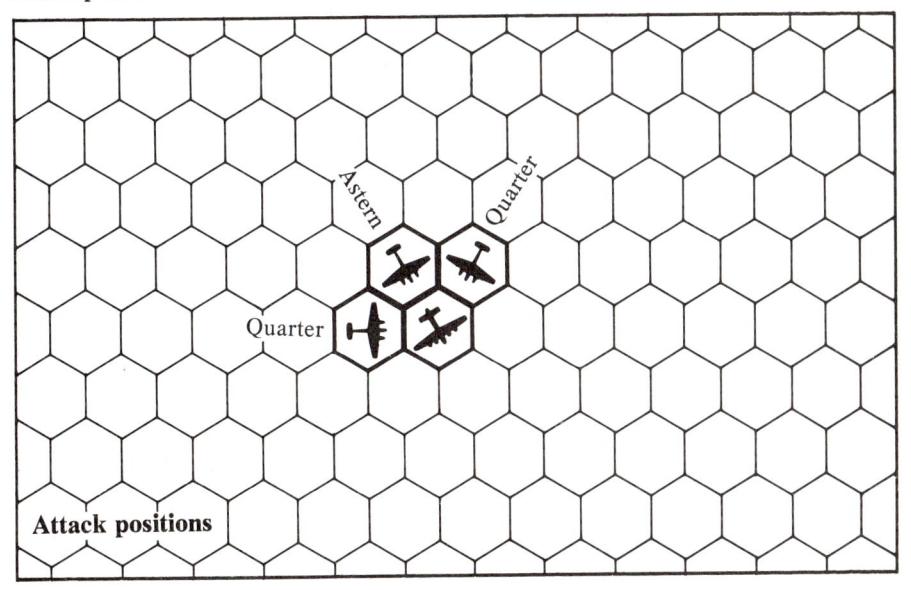

Attack positions

Finally, we have the interception. This is a little less than realistic as for the good of the game we have to keep the pot boiling, and this is precisely where too much realism would spoil it. In theory, the bomber can shoot back. In practice, less than one in five bombers showed any awareness that an interceptor was in the vicinity until they were actually shot at. I therefore prefer to discount the bomber's defensive fire entirely.

To make a successful attack, it is necessary to slow down. An attack can only be made if the fighter is travelling at the same speed, or not more than one hex faster than its target. This means that an interceptor has to slow down to three hexes per move to attack a bomber, but itself becomes vulnerable to an intruder sneaking up behind at four hexes per move. There are three possible hexes from which an attack can be made; the hex immediately astern of the target, and the hexes to either side of this, which count as quarter attacks. Head-on or beam attacks are not possible at night. Firing is carried out with one normal (six spot) dice. A score of four, five or six from a stern attack means that the target is shot down, but from the quarter, only a score of six will suffice.

If you wish to have a competiton rather than just play the game for the fun of it, you will need a points system of scoring. Working on the size of game suggested, with 18 bombers, four interceptors and three intruders, the RAF player scores one point for each bomber going off the board on the far side successfully, and five points for each interceptor shot down. The Luftwaffe player scores three points for each bomber shot down and nothing for intruders. This is not to say that shooting down intruders is not worthwhile; each intruder destroyed makes the job of knocking down the bombers so much easier, therefore the value of swatting an intruder is measured by how many extra bombers you can get. It sounds simple; it is simple. Just try it.

Select bibliography

Bekker, Cajus: *The Luftwaffe War Diaries* (Macdonald, 1967).
Brown, David: *Carrier Fighters* (Macdonald & Jane's, 1975).
Chamberlain, Peter, and Gander, Terry: *Anti-aircraft guns* (Macdonald & Jane's WW2 Fact File, 1975).
Dempster, Derek, and Wood, Derek: *The Narrow Margin* (Hutchinson, 1961).
Gunston, Bill: *Night Fighters* (Patrick Stephens, 1976).
Lloyd, H.P.: *Briefed to Attack* (Hodder & Stoughton, 1949).
Price, Alfred: *Aircraft versus Submarine* (Kimber, 1973); *Battle Over the Reich* (Ian Allan, 1973); *World War II Fighter Conflict; The Bomber in World War II; Instruments of Darkness* (Macdonald & Jane's, 1975, 1976, 1977).
Shores, Christopher: *Pictorial History of the Mediterranean Air War* (Ian Allan, 1974); *Ground Attack Aircraft of World War II* (Macdonald & Jane's, 1977).
Sims, Edward H.: *The Fighter Pilots* (Cassel, 1967).
Townsend, Peter: *Duel of Eagles* (Weidenfeld, 1970).